Maria Mironova

Search for Higgs Boson Decays to Charm Quarks with the ATLAS Experiment and Development of Novel Silicon Pixel Detectors

Doctoral Thesis accepted by
University of Oxford, Oxford, UK

Springer

Author
Dr. Maria Mironova
Lawrence Berkeley National Laboratory
Berkeley, CA, USA

Supervisor
Prof. Daniela Bortoletto
University of Oxford
Oxford, UK

ISSN 2190-5053 ISSN 2190-5061 (electronic)
Springer Theses
ISBN 978-3-031-36219-4 ISBN 978-3-031-36220-0 (eBook)
https://doi.org/10.1007/978-3-031-36220-0

This Springer imprint is published by the registered company Springer Nature Switzerland AG
The registered company address is: Gewerbestrasse 11, 6330 Cham, Switzerland

Supervisor's Foreword

The discovery of the Higgs boson, which occurred just over 10 years ago, was pivotal for particle physics since the observation of this long-sought particle also provided direct evidence for the existence of the Higgs field, which plays a unique role in the Standard Model (SM) of elementary particle physics. The masses of W and Z bosons are generated by spontaneous symmetry breaking of the electroweak gauge symmetry due to the Higgs field. The masses of fundamental fermions, quarks and leptons, are related to the interaction between the Higgs field and massless fermion fields, described by the Yukawa couplings. The differences in the Yukawa couplings of different quarks and leptons are at the core of the mystery of the mass hierarchy between different generations of fermions. The Yukawa interactions play a fundamental role in our universe. For example, the electron's Yukawa coupling sets the electron mass and, therefore, the atoms' size and energy levels. The Yukawa couplings of the first generation of quarks explain the stability of protons because the down quark is heavier than the up quark. We have conclusively established that the Higgs mechanism is responsible for the mass of the third generation of charged fermions (tau lepton, top and bottom quark). Evidence for the decay of the Higgs boson to a second-generation lepton, the muon, has been recently reported at the LHC. Nonetheless, until we observe the Higgs Boson decay to a second-generation quark, our understanding of the role of the Higgs mechanism in giving mass to matter particles remains incomplete.

In this thesis, Dr. Mironova has conducted an insightful study of the coupling of the Higgs to charm, which while small, is the most promising decay mode to probe that the Yukawa hypothesis also works for the second-generation quarks. She searched for the associated production of a Higgs boson H with an electroweak boson (V), followed by the decay $H \rightarrow c\bar{c}$, and set powerful constraints on the Higgs-charm coupling modifier to be less than 8.5 times what is expected in the Standard Model. Furthermore, by combining her analysis with the measurement of $VH(H \rightarrow b\bar{b})$, she established at 95% CL that the Higgs-charm couplings are smaller than the Higgs-bottom Yukawa couplings. This critical step paves the way for even better future measurements through analysis improvements and larger datasets, which will become available in the high luminosity phase of the Large Hadron Collider, the HL-LHC, or future colliders.

Dr. Mironova also contributed superbly to developing the ATLAS ITk pixel detector, which is currently under construction and will be deployed at the heart of the experiment for the HL-LHC. The readout chip for ITk Pixel, ITkPixV1, is the first 65 nm readout chip used in a particle physics experiment. It contains one billion transistors and features a 5-Gbit connection for high hit intensities. A low-power, low-noise analogue front-end ensures high readout speeds and low detection thresholds. The readout chips will operate in the very challenging environments of the HL-LHC, where they will receive total ionising doses of up to 1 Grad. In her thesis, using an X-ray system that she set up, Dr. Mironova has shown that the ITkPixV1 chip is as radiation tolerant as expected and should survive the intended lifetime in the ITk upgrade. Finally, the thesis describes monolithic pixel detectors as a possible technology for future pixel detectors and summarises their characterisation with an X-ray microbeam at the Diamond Light source.

The physics analysis and the detector work are presented beautifully, making this thesis an essential resource for beginning students and experts interested in the very challenging analysis and silicon detectors.

Oxford, UK
May 2023

Prof. Daniela Bortoletto

Abstract

The coupling of the Higgs boson to the third generation of fermions and heavy vector bosons has been experimentally well established at the Large Hadron Collider (LHC), and the measured properties agree well with Standard Model (SM) expectation. Evidence of the Higgs coupling to second-generation quarks is still elusive, and measurements of Higgs boson decays to *c*-quarks are the next most promising decay mode to probe the validity of the SM.

This thesis describes a search for $H \rightarrow c\bar{c}$ decays in the *VH* production mode, using a *pp* collision dataset with an integrated luminosity of 139 fb^{-1} at a centre-of-mass energy of 13 TeV, collected with the ATLAS detector. The analysis includes three different channels based on the decay of the vector boson and uses dedicated machine learning algorithms to identify *c*-quarks in the experiment. The observed (expected) upper limit for the $VH(H \rightarrow c\bar{c})$ process is $26(31^{+12}_{-8})$ times the predicted standard model cross-section times branching fraction, at the 95% confidence level, and a direct constraint on the Higgs charm coupling modifier of $|k_c| < 8.5$ is set. In a combination of this analysis with a measurement of the $VH(H \rightarrow b\bar{b})$ process an upper limit of $|k_c/k_b| < 4.5$ is determined.

Significant analysis improvements and a larger dataset are needed to observe $H \rightarrow c\bar{c}$ decays in the future. At the High-Luminosity LHC (HL-LHC), the ATLAS detector will collect at least 3000 fb^{-1} of integrated luminosity over 10 years, which requires significant improvements to the detector to cope with increased radiation damage and higher data rates. This thesis discusses the characterisation of the ATLAS pixel detector readout chip for the inner detector upgrade (ITk) for HL-LHC. The radiation tolerance of the chip is tested in irradiations using X-rays and protons. All readout chip components, such as the digital logic and analogue front-end, are sufficiently radiation-tolerant to withstand the 1 Grad total ionising dose expected at HL-LHC.

Finally, the thesis describes monolithic pixel detectors as a possible technology for future pixel detectors and summarises the characterisation of a monolithic detector prototype in an X-ray beam test.

Preface

The work presented in this thesis has been completed within the ATLAS Collaboration and in close collaboration with colleagues. This thesis aims to outline the main results of this work, focusing on my own contributions. Figures and results obtained by others are clearly labelled as such. All other figures are either the result of collaboration with others (marked with an "ATLAS" label) or my own. Outlined below are the main projects I worked on during my D.Phil and my specific contributions.

Search for Higgs boson decays to charm quarks I was one of the main analysers in the recent ATLAS search for Higgs boson decays to charm quarks in the VH production mode. The results are described in Chap. 4. In particular:

- I studied the truth-tagging method, which was used to improve the available statistics of the simulated samples.
- I introduced a simultaneous measurement of well-known Standard Model processes, the diboson signals $VW(W \rightarrow cq)$ and $VZ(Z \rightarrow c\bar{c})$, to cross-check the analysis strategy.
- I studied and defined several control regions in the fit to constrain the $t\bar{t}$ background in the *0-lepton* and *1-lepton* channels, and the V+ jets background in all channels. I also derived signal and background modelling uncertainties for all processes considered in the analysis. I defined the final analysis model in terms of analysis categories and treatment of systematic uncertainties.
- I performed the final statistical analysis for extracting the parameters of interest and conducted many studies to understand the results.
- I worked on the statistical combination of the $H \rightarrow c\bar{c}$ and $H \rightarrow b\bar{b}$ analyses, measuring the Higgs coupling to b- and c-quarks simultaneously. In particular, I performed a diboson cross-check combination and worked on fit diagnostics for the combined analyses.

Prospects for the $VH(H \to b\bar{b}/c\bar{c})$ analyses at HL-LHC I studied the expected sensitivity of the ATLAS measurements of Higgs decays to c- and b-quarks with the increased dataset and improved detector performance expected at HL-LHC. In particular, I evaluated the impact of c-quark identification efficiencies on the expected sensitivity to $H \to c\bar{c}$ decays at HL-LHC. I also estimated the effect of the MET trigger threshold on the $H \to c\bar{c}$ and $H \to b\bar{b}$ analyses to provide input into the decision on the trigger menu in ATLAS. Finally, I produced results and was the primary author of a PUB note on the extrapolation of the $H \to c\bar{c}$ and $H \to b\bar{b}$ analyses to HL-LHC as input for the US Snowmass Particle Physics Community Planning Exercise.

Radiation tolerance of ITkPixV1 chips In collaboration with Berkeley Lab, I performed many tests of the latest ATLAS pixel readout chip prototype for the upgrade to HL-LHC to verify it can withstand the required levels of radiation. The results are described in Chap. 6. In particular:

- I commissioned and calibrated the X-ray system and irradiation set-up in Oxford.
- I performed 15 irradiation campaigns studying the radiation tolerance of the latest ATLAS pixel readout chip (ITkPixV1), providing information about the expected behaviour of the digital and analogue front-end at the high total ionising doses expected at HL-LHC.
- I conducted multiple irradiation campaigns to understand the impact of low-dose rate effects on small transistors.
- I set up and performed an irradiation campaign of ITkPixV1 pixel modules using protons at the Birmingham Cyclotron to study the impact of radiation on both pixel sensors and readout chips.

X-ray testbeam of mini-MALTA chips With several collaborators, I conducted an X-ray testbeam campaign at Diamond Light Source testing the pixel response of a monolithic detector prototype (miniMALTA)—a candidate for a future upgrade of the ATLAS detector and future colliders. I was the primary analyser of the data and author of the two subsequent papers, and developed a new method for estimating pixel efficiency from photon response data. This work is described in Chap. 7.

Testing of sensors and modules for ATLAS ITk I was the primary person responsible for testing ATLAS ITk sensors and modules at the Oxford Physics Microstructure Detector (OMPD) lab. During my ATLAS qualification task, I characterised pixel sensors from different vendors before and after irradiation for the ATLAS Market Survey to identify potential suppliers of pixel sensors for the ITk upgrade. I set up several readout systems for ATLAS pixel modules used for testing and quality control during production. I was responsible for performing the quality control measurements for the RD53A quad modules assembled at Oxford and worked on thermal stress studies on RD53A dual-chip modules. This work is not described in this thesis.

ITk Simulation I contributed to the simulation of the ITk detector, used in many physics and performance studies for the upgrade. In particular, I worked on the migration of the pixel detector geometry to a different, more maintainable framework (`GeoModelXML`). This work is not described in this thesis.

Berkeley, USA Maria Mironova

Acknowledgements

Thank you very much to everyone who contributed to the completion of this thesis and to everyone I had the pleasure of collaborating with during my D.Phil.

First and foremost, I would like to thank my supervisor Daniela Bortoletto for making this D.Phil possible. I am deeply grateful for your guidance and support during the last 4 years and your advice on physics and beyond. Thank you very much for helping me expand my research interests and giving me the possibility to pursue new projects—you have taught me a lot, and it has been a privilege to be your student!

Thanks to the $H \rightarrow b\bar{b}$ group for providing a great environment to work in and the $VH(H \rightarrow c\bar{c})$ analysis team—it has truly been a pleasure to work with you. In particular, I am very thankful to Andy Chisholm, Tristan du Pree and Hannah Arnold for guiding the analysis to completion and the many fruitful discussions. Many thanks to Marko Stamenkovic for the help, conversations and friendship.

One of the favourite parts of my D.Phil were my days spent working in the OPMD lab—thank you to Daniela and Ian for setting up such a fantastic laboratory. I am very thankful to Richard Plackett for ensuring the lab runs smoothly and for all of the help and guidance with my lab work. A special thanks to my favourite lab people, Kaloyan Metodiev and Dan Wood; working with you made my time in the lab even more enjoyable. Thank you also to Alex Knight, Dan Weatherill, Gale Lockwood and Kirk Arndt.

I am very grateful for the collaboration with LBNL during my D.Phil. In particular, I want to thank Maurice Garcia-Sciveres and Timon Heim for the remote supervision with the ITkPixV1 studies—I have learnt a lot from you and am looking forward to working even more closely with you in the future! Thanks also to Aleksandra Dimitrevska, Daniel Antrim and Hongtao Yang.

Thank you to the MALTA group at CERN for making the Diamond beam test possible and for the smooth collaboration, especially thanks to Carlos Solans Sanchez, Valerio Dao and Abhishek Sharma.

Thank you to everyone in the Bortoletto-Shipsey group—I have greatly enjoyed our meetings and the productive conversations. A special thanks to Ian Shipsey for many interesting discussions and valuable guidance. Thank you to Cecilia Tosciri, Luca Ambroz and Luigi Vigani—your advice during my first year was much appreciated.

Thank you to the ATLAS group in Göttingen, and my former supervisor Arnulf Quadt, for introducing me to Particle Physics.

I am deeply thankful to the Science and Technology Facilities Council and the Oxford Scatcherd European Scholarship for financially supporting my D.Phil. Thank you also to the German Academic Scholarship Foundation for supporting me during my undergraduate studies.

A big thanks to the Particle Physics administration team: Sue Geddes, Jennifer Matthews and Kim Proudfoot—your help, professionalism and kindness were invaluable.

I am grateful to my friends during my time in Oxford, especially James Grundy and Luke Scantlebury-Smead. A special thanks to Hayden Smith—I am extremely thankful for your friendship and very happy we got to experience the D.Phil together.

Finally, I am incredibly grateful to my parents, Lidia and Sergey, for always being there and taking care of me. This would not be possible without your unwavering love and support—thank you!

Contents

Chapter 1
Introduction

The Standard Model of Particle Physics (SM) has been a hugely successful theory in describing our current best understanding of the fundamental properties of the universe [1, 2]. It is a theoretical formalism driven by the concepts of symmetries and gauge invariance, and it provides a description of the electromagnetic, strong and weak interactions. The theory is completed by introducing spontaneous symmetry breaking and the Higgs mechanism, which allows for the existence of massive bosons. The SM has been extensively probed experimentally and has shown good agreement of experimental observations with theoretical predictions.

In 2012, the Higgs boson discovery by the ATLAS and CMS collaborations at the Large Hadron Collider (LHC) was a major success in confirming the validity of the Standard Model of Particle Physics [3, 4]. Since then, the new particle has been studied in increasing detail, confirming its coupling to many of the massive particles in the SM and showing properties consistent with the SM expectation. Nevertheless, several Higgs decay modes have not been observed yet, and the measurement of these is crucial to understanding the behaviour of the Higgs boson completely. In addition, there is significant evidence that the SM does not fully describe all physical observations in nature, and physics beyond the SM is necessary. For instance, the existence of dark matter is well-known from astrophysical observations, but it is not included in the SM. Additionally, the nature of neutrinos is not well-understood, and neutrino masses are currently not included in the SM. The exact nature of the Higgs mechanism also needs to be understood further. For example, the origin of the different particle masses and couplings to the Higgs is not clear. Additionally, we do not know why the measured Higgs mass is so much smaller than its predicted value. Therefore, the measurement of Higgs boson properties remains a focus of the LHC physics programme.

This thesis presents the results of a recent search for Higgs boson decays into charm quarks with the ATLAS detector, using a pp collision dataset of 139 fb^{-1} at a center-of-mass energy of 13 TeV. This search aims to measure as precisely as possible the coupling of the Higgs boson to the second generation of quarks. This coupling

M. Mironova, *Search for Higgs Boson Decays to Charm Quarks with the ATLAS Experiment and Development of Novel Silicon Pixel Detectors*, Springer Theses,
https://doi.org/10.1007/978-3-031-36220-0_1

has not been directly measured yet, and a deviation of the measured coupling from the SM expectation would provide clear evidence for new physics, as predicted in several scenarios [5–11]. In the search, an upper limit on the $VH(H \rightarrow c\bar{c})$ signal strength is determined, as well as a direct measurement of the Higgs-charm coupling κ_c. A statistical combination with the measurement of $VH(H \rightarrow b\bar{b})$ decays also provides the opportunity to constrain the ratio κ_c/κ_b.

Experimental measurements are only as good as the detectors used to collect the data. Consequently, it is crucial to constantly improve our detector capabilities to collect more data more rapidly and enhance the precision of the measurements. Around 2028, the ATLAS detector is expected to upgrade its inner tracking system in preparation for the High-Luminosity LHC (HL-LHC), where a significantly larger dataset of at least 3000 fb^{-1} will be collected. This all-silicon inner tracker (ITk) is needed to cope with the higher data rates and expected radiation damage at HL-LHC. It will also improve the reconstruction of particles in the experiment, which will enhance analyses such as the $VH(H \rightarrow c\bar{c})$ search. This thesis also describes efforts undertaken as part of the ITk pixel detector upgrade. In particular, the radiation tolerance of the pixel readout chip is studied in detail to ensure an operational detector throughout the lifetime of ITk. Finally, a brief outlook is given on the potential of new pixel detector technologies, such as monolithic detectors, which can further improve measurements at the LHC and future collider experiments.

The thesis is structured as follows. Chapter 2 provides a brief overview of the theoretical formulation of the Standard Model of Particle Physics and the properties and methods related to measurements of the Higgs boson at the LHC. Chapter 3 includes a description of the ATLAS detector and the methods for reconstructing particles in the experiment. Chapter 4 describes the $VH(H \rightarrow c\bar{c})$ analysis using a pp collision dataset of 139 fb^{-1}, and the extrapolation of this analysis to HL-LHC. Chapter 5 provides a theoretical overview of silicon pixel detectors. Chapter 6 discusses the investigations into the radiation tolerance of the pixel readout chip for the ATLAS ITk upgrade in measurements with X-rays and protons. Chapter 7 outlines the possible benefits of using monolithic pixel detectors and describes an X-ray beam test performed at Diamond Light Source. Chapter 8 provides a brief summary and conclusions.

References

1. Thomson M (2013) Modern particle physics. Cambridge University Press, New York. ISBN: 978-1-107-03426-6
2. Perkins DH (1982) Introduction to high energy physics. ISBN: 978-0-521-62196-0
3. ATLAS Collaboration (2012) Observation of a new particle in the search for the StandardModel Higgs boson with the ATLAS detector at the LHC. Phys Lett B 716:1. https://doi.org/10.1016/j.physletb.2012.08.020. arXiv:1207.7214 [hep-ex]
4. CMSCollaboration (2012) Observation of a new boson at a mass of 125GeV with the CMS experiment at the LHC. Phys Lett B 716:30. https://doi.org/10.1016/j.physletb.2012.08.021. arXiv:1207.7235 [hep-ex]

5. Delaunay C et al (2014) Enhanced Higgs boson coupling to charm pairs. Phys Rev D 89:033014. https://doi.org/10.1103/PhysRevD.89.033014
6. Perez G et al (2015) Constraining the charm Yukawa and Higgs-quark coupling universality. Phys Rev D 92:033016. https://doi.org/10.1103/PhysRevD.92.033016
7. Botella FJ et al (2016) What if the masses of the first two quark families are not generated by the standard model Higgs boson? Phys Rev D 94(11):115031. https://doi.org/10.1103/PhysRevD.94.115031. arXiv:1602.08011 [hep-ph]
8. Bar-Shalom S, Soni A (2018) Universally enhanced light-quarks Yukawa couplings paradigm. Phys Rev D 98:055001. https://doi.org/10.1103/PhysRevD.98.055001
9. Ghosh D, Gupta RS, Perez G (2016) Is the Higgs mechanism of fermion mass generation a fact? A Yukawa-less first-two-generation model. Phys Lett B 755:504–508. ISSN: 0370-2693. https://doi.org/10.1016/j.physletb.2016.02.059
10. Egana-Ugrinovic D, Homiller S, Meade PR (2019) Aligned and spontaneous flavor violation. Phys Rev Lett 123(3):031802. https://doi.org/10.1103/PhysRevLett.123.031802. arXiv:1811.00017 [hep-ph]
11. Egana-Ugrinovic D, Homiller S, Meade PR (2019) Higgs bosons with large couplings to light quarks. Phys Rev D 100(11):115041. https://doi.org/10.1103/PhysRevD.100.115041. arXiv:1908.11376 [hep-ph]

Chapter 2
Theoretical Background

Our current best understanding of the fundamental particles of the Universe is described in the Standard Model of Particle Physics (SM) [1, 2]. This chapter outlines the structure of the Standard Model, the formalism that leads to the prediction of the Higgs boson, the phenomenology related to Higgs boson measurements at the LHC, and how theoretical predictions are made for collider experiments.

2.1 The Standard Model of Particle Physics

The particles of the Standard Model are illustrated in Fig. 2.1. They can be classified by their intrinsic angular momentum, i.e. their *spin*. Particles with half-integer spin follow Fermi-Dirac statistics and are known as *fermions*, whereas particles with integer spin obey Bose-Einstein statistics and are known as *bosons*. Fermions exist in three *generations*, where the first generation particles are the constituents of all stable matter, and the second and third-generation fermions can be seen as heavier replicas of the first-generation fermions. The second and third-generation charged fermions are not stable and eventually decay into their first-generation counterparts. The fermions can be categorised further by considering their quantum numbers. Fermions that carry colour charge are known as *quarks*. *Up-type* quarks have an electrical charge of $+\frac{2}{3}|e|$ and exist in three flavours: up (u), charm (c) and top (t). Similarly, *down-type* quarks carry a charge of $-\frac{1}{3}|e|$ and can be identified in three flavours: down (d), strange (s) and bottom (b). Next, *leptons* are fermions that have no colour charge and can have an integer electric charge of $-|e|$ or no electrical charge. The charged leptons are the electron (e), muon (μ) and tau-lepton (τ), and each of them has a neutral counterpart, referred to as a *neutrino* and denoted as ν_e, ν_μ and ν_τ.

The fundamental particles interact via the four fundamental forces, which describe all interactions observed in nature: the strong, electromagnetic and weak forces, as

M. Mironova, *Search for Higgs Boson Decays to Charm Quarks with the ATLAS Experiment and Development of Novel Silicon Pixel Detectors*, Springer Theses,
https://doi.org/10.1007/978-3-031-36220-0_2

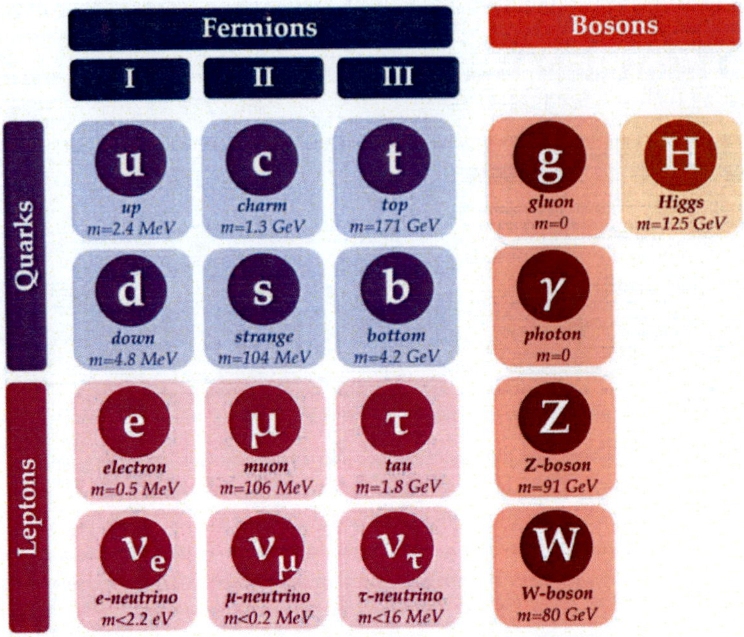

Fig. 2.1 The particles of the Standard Model of Particle Physics. Data from [3]

well as gravity. The effect of gravity is negligible for fundamental particles, so it is not included in the formulation of the SM. The other three forces in the SM are mediated by spin-1 particles, known as *gauge bosons*. The mediator of the electromagnetic force is the photon (γ), which is massless and has no charge. The strong force is mediated by the gluon (g), which is massless, but carries colour charge [4]. Finally, the mediators of the weak force are the massive charged $W^{\pm}$ and neutral Z boson (often just denoted as W and Z) [5]. The last constituent of the SM is the Higgs boson (H), a spin-0 massive boson, emerging from the Higgs field that gives rise to particle masses in the SM.

The Standard Model is a quantum field theory (QFT) that describes particles as perturbations of quantum fields that satisfy particular field equations. The theory is constructed upon the Lagrangian formalism and generalised from the theory of *Quantum Electrodynamics (QED)*. The Lagrangian density $\mathscr{L} = \mathscr{L}(\psi(x), \partial_\mu \psi(x))$ describes the dynamics of a field $\psi(x)$. Global and local symmetries play an important role in this formalism. A global symmetry means that a Lagrangian density is invariant under a transformation that is identical for all space-time points. In contrast, the transformation can differ as a function of space-time for a local symmetry. Symmetries are important, as each symmetry gives rise to a conserved quantity, as stated by Noether's theorem [6].

The formulation of the Standard Model starts by considering the dynamics of free fermions, which are governed by the Dirac equation of quantum mechanics

$$(i\gamma^\mu \partial_\mu - m)\psi = 0. \tag{2.1}$$

Here, $\psi = \psi(x)$ is the field spinor, γ^μ represents the Dirac γ-matrices, and m is the mass of the fermion. The Dirac equation has four solutions, two of which imply the existence of negative energies. These negative energy solutions are interpreted as *anti-particles*, i.e. positive energy particles propagating backwards in time, which have the same mass, but the opposite charge of particles. The Dirac equation corresponds to a Lagrangian density of

$$\mathscr{L}_{\text{Dirac}} = \bar{\psi}(i\gamma^\mu \partial_\mu - m)\psi. \tag{2.2}$$

Now, the Lagrangian is required to be invariant under a $U(1)$ local transformation, a requirement also known as gauge invariance. The transformation can be written as

$$\psi \rightarrow \psi' = e^{iq\alpha(x)}\psi, \tag{2.3}$$

where $e^{iq\alpha(x)}$ is a complex space-time dependent phase. To ensure the gauge invariance of the Lagrangian, the partial derivative ∂_μ can be replaced with the *covariant derivative* D_μ:

$$\partial_\mu \rightarrow D_\mu = \partial_\mu + iqA_\mu, \tag{2.4}$$

where A_μ is a vector field, which transforms under $U(1)$ as

$$A_\mu \rightarrow A'_\mu = A_\mu - \partial_\mu \alpha. \tag{2.5}$$

The A_μ field can be interpreted as the gauge field for the electromagnetic interaction, i.e. the massless photon, with the interaction strength q. Generally, the invariance of a Lagrangian under a gauge transformation results in the introduction of spin-1 particles (gauge bosons) in this theory, which serve as the mediator of a particular interaction. In this formalism, mass terms of the form $m^2 A^\mu A_\mu$ are not gauge invariant and the mediators have to be massless. With the above requirements on gauge invariance under $U(1)$ transformations, the QED Lagrangian can now be written as

$$\begin{aligned} \mathscr{L}_{\text{QED}} &= \bar{\psi}(i\gamma^\mu \partial_\mu - m)\psi + q\bar{\psi}\gamma^\mu A_\mu \psi - \frac{1}{4}F_{\mu\nu}F^{\mu\nu} \\ &= \bar{\psi}(i\gamma^\mu D_\mu - m)\psi - \frac{1}{4}F_{\mu\nu}F^{\mu\nu} \end{aligned} \tag{2.6}$$

where $F_{\mu\nu} = \partial_\mu A_\nu - \partial_\nu A_\mu$ is the electromagnetic field tensor, $\frac{1}{4}F_{\mu\nu}F^{\mu\nu}$ is the kinetic term of the gauge field and the $q\bar{\psi}\gamma^\mu A_\mu \psi$ term describes the interaction between the fermion and the gauge boson.

The mechanism described above can be extended to other symmetry groups. In particular, it is also possible to consider non-abelian symmetry groups, i.e. groups whose generators do not commute. An example of this is $SU(3)$, which describes the strong interaction in the *Quantum Chromodynamics (QCD)* theory. For the strong interaction, the conserved quantum number is the colour charge. Only quarks have a colour charge in the SM and can have colour charges *red*, *blue* or *green*. The QCD Lagrangian is:

$$\mathscr{L}_{\text{QCD}} = \sum_q \bar{\psi}_q (i\gamma^\mu D_\mu - m_q)\psi_q - \frac{1}{4} G^a_{\mu\nu} G^{\mu\nu}_a, \tag{2.7}$$

where q corresponds to the sum over the quarks. Similar to QED, a covariant derivative can be defined to ensure invariance under $SU(3)$ transformations:

$$D_\mu = \partial_\mu + i g_s \frac{\lambda^a}{2} G^a_\mu. \tag{2.8}$$

Here g_s is the coupling constant of the strong interaction, which is often expressed in terms of $\alpha_s = \frac{g_s}{4\pi}$. The Gell-Mann matrices, λ^a, are a set of matrices that generate the $SU(3)$ group, and G^a_μ corresponds to the gluons, the eight fields that mediate the strong interaction. From these fields, the strong interaction tensor can be defined as:

$$G^a_{\mu\nu} = \partial_\mu G^a_\nu - \partial_\nu G^a_\mu - g_s f^{abc} G^b_\mu G^c_\nu, \tag{2.9}$$

where f^{abc} are the structure constants of the $SU(3)$ group. As can be seen from the structure of the strong interaction tensor, the gluons themselves carry colour charges and can therefore interact with each other. This leads to a phenomenon known as *colour confinement* [7–9]. Quarks and gluons can only exist in neutral colour states. This means that gluons do not propagate macroscopic distances, and quarks can only exist in bound states known as *hadrons*. These hadrons can either consist of a quark and an anti-quark, known as a *meson*, or as a colour-neutral combination of three quarks, known as a *baryon*. The bound states are produced in a process known as *hadronisation*.

A further complication in QCD calculations is that the strong coupling strength α_s strongly depends on the energy at which it is measured. It is large at small energies (<1 GeV) and decreases as the energy increases. This behaviour is known as *asymptotic freedom*. At energies above around 100 GeV, α_s is sufficiently small to allow perturbative calculations in the derivation of QCD predictions. However, higher-order corrections are needed for computations to converge. Some of these terms, e.g. quark self-energy corrections via loops, can lead to infinities in the calculations. This feature is known as *ultraviolet divergences* and is fixed by renormalising fields and parameters to absorb these divergences. The corrections depend on arbitrary scale parameters, known as the *renormalisation scale* μ_R and the *factorisation scale* μ_F [10, 11].

The second non-abelian symmetry group relevant to the Standard Model is $SU(2) \times U(1)$, which unifies the electromagnetic and weak forces in the SM. A local $SU(2)$ transformation is defined as

$$\psi \to \psi' = e^{ig\alpha^a(x)\frac{\sigma^a}{2}}\psi, \tag{2.10}$$

where σ^a are the Pauli matrices, which are the generators of the $SU(2)$ group. Similarly to the previous two cases, a covariant derivative can be defined to ensure gauge invariance:

$$D_\mu = \partial_\mu + ig\frac{\sigma_a}{2}W_\mu^a \tag{2.11}$$

Here, three gauge fields W_μ^1, W_μ^2 and W_μ^3 are introduced, which have an interaction strength g. In the SM, the weak interaction is described by the $SU(2)_L$ group, where the subscript L indicates that the charged-current interaction only couples to left-handed (LH) particles states (and right-handed (RH) antiparticle states). Fermionic fields are split into their left- and right-handed components using the projection operators P_L and P_R, as

$$\psi = P_L\psi + P_R\psi = \frac{1-\gamma^5}{2}\psi + \frac{1+\gamma^5}{2}\psi = \psi_L + \psi_R. \tag{2.12}$$

For the weak interaction, the *weak isospin* charge is defined, and isospin doublets represent the LH and RH particles. LH particles have a weak isospin of $I_W = \frac{1}{2}$, and a third component of the weak isospin of $I_W^3 = \pm\frac{1}{2}$. For RH particles, $I_W = 0$ and $I_W^3 = 0$, which means that the W_μ^a gauge bosons only couple to the LH particles. The physical weak charged-current interactions, which correspond to the $W^\pm$ bosons, are described as linear combinations of these fields as

$$W_\mu^\pm = \frac{1}{\sqrt{2}}(W_\mu^1 \mp iW_\mu^2). \tag{2.13}$$

Experimentally it has indeed been observed that the W boson only couples to LH particles. However, the Z boson has been observed to couple to both left- and right-handed particles. Therefore, an additional symmetry group $U(1)_Y$ is introduced to accommodate this observation in the SM formalism. The corresponding charge is the *weak hypercharge* Y, and an additional gauge field B_μ is introduced, with coupling g'. The weak hypercharge is defined as

$$Y = 2(Q - I_W^3), \tag{2.14}$$

where Q is the electromagnetic charge. For the $SU(2) \times U(1)$ group, the covariant derivative can then be defined as

$$D_\mu = \partial_\mu + ig\frac{\sigma_a}{2}W_\mu^a + ig'\frac{Y}{2}B_\mu. \tag{2.15}$$

In this Glashow-Salam-Weinberg (GSW) model [12–14], the physical electromagnetic photon A_μ and the Z-boson Z_μ result from the linear combinations of the B_μ and W^3_μ fields, via the *weak mixing angle* θ_W

$$\begin{pmatrix} A_\mu \\ Z_\mu \end{pmatrix} = \begin{pmatrix} \cos\theta_W & \sin\theta_W \\ -\sin\theta_W & \cos\theta_W \end{pmatrix} \begin{pmatrix} B_\mu \\ W^3_\mu \end{pmatrix}, \tag{2.16}$$

and the coupling strengths of the weak interaction g_W and g_Z, and electromagnetic interactions e are related as:

$$e = g_W \sin\theta_W = g_Z \sin\theta_W \cos\theta_W. \tag{2.17}$$

Combining the description of interactions, the SM unifies the electroweak and QCD theories under the group $SU(3)_C \times SU(2)_L \times U(1)_Y$. However, the local gauge invariance requirement can only be satisfied if all the bosons are massless. In addition, the split of the fermionic fields into left- and right-handed components breaks the gauge invariance of the mass term in the electroweak Lagrangian. Experimentally the fermions and the W and Z bosons have been observed to have mass. Therefore, an additional mechanism needs to be included in the electroweak sector of the SM, which spontaneously breaks the symmetry and gives rise to particle masses: the *Brout-Englert-Higgs mechanism* [15–18].

2.2 The Higgs Mechanism

The Higgs mechanism describes how the gauge bosons acquire their mass through interactions with the Higgs field. This new field is described as a weak isospin doublet, with one neutral field ϕ^0 and one charged field ϕ^+, which can be expressed as complex scalar fields with four degrees of freedom:

$$\phi = \begin{pmatrix} \phi^+ \\ \phi^0 \end{pmatrix} = \frac{1}{\sqrt{2}} \begin{pmatrix} \phi_1 + i\phi_2 \\ \phi_3 + i\phi_4 \end{pmatrix} \tag{2.18}$$

The corresponding Lagrangian is

$$\mathscr{L} = (D_\mu\phi)^\dagger(D^\mu\phi) - V(\phi), \tag{2.19}$$

where the first term describes the kinetic term of the field, and D_μ is the covariant derivative defined in Eq. 2.15. The second term $V(\phi)$ is the *Higgs potential*:

$$V(\phi) = \mu^2\phi^\dagger\phi + \lambda(\phi^\dagger\phi)^2, \tag{2.20}$$

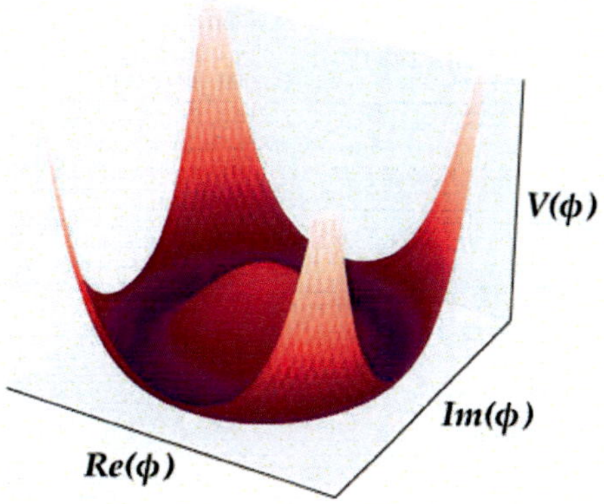

Fig. 2.2 Illustration of the expected Higgs potential for $\mu^2 < 0$

where μ and λ are scalar constants. The first term of the potential is the mass term, while the second term describes the self-interaction of the field. The minimum of the potential corresponds to the vacuum state. If $\mu^2 > 0$, the Higgs potential has one unique minimum at $\phi = 0$. However, if $\mu^2 < 0$, a set of degenerate minima can be found at a circumference of

$$(\phi^\dagger \phi) = \frac{1}{2}(\phi_1^2 + \phi_2^2 + \phi_3^2 + \phi_4^2) = \frac{v^2}{2} = \frac{|\mu|^2}{2\lambda}, \tag{2.21}$$

which is illustrated in Fig. 2.2. The parameter $v = \sqrt{\frac{|\mu|^2}{\lambda}}$ is known as the *vacuum expectation value* of the field ϕ. All of the available minima are equivalent and connected by an $SU(2)$ transformation. The choice of the vacuum state breaks the symmetry of the Lagrangian in a mechanism known as *spontaneous symmetry breaking*. One can choose the vacuum state where $\phi_1 = \phi_2 = \phi_4 = 0$ and $\phi_3 = v$, which leads to a vacuum expectation of

$$\langle 0|\phi|0\rangle = \frac{1}{\sqrt{2}}\begin{pmatrix} 0 \\ v \end{pmatrix}. \tag{2.22}$$

It is now possible to expand around this vacuum expectation state to quantify the excitation of the field around the vacuum:

$$\phi(x) = \frac{1}{\sqrt{2}}\begin{pmatrix} \phi_1(x) + i\phi_2(x) \\ v + i\eta(x) + i\phi_4(x) \end{pmatrix}. \tag{2.23}$$

In this parametrisation, the field ϕ is described by one real physical field $\eta(x)$ and three unphysical fields $\phi_{1,2,4}$, which are known as the *Goldstone fields* [17]. It is possible to choose a particular gauge, known as the *unitary gauge*, where the vector fields absorb the Goldstone fields. In this case, the expansion around to vacuum simplifies to

$$\phi(x) = \frac{1}{\sqrt{2}} \begin{pmatrix} 0 \\ v + h(x) \end{pmatrix}. \tag{2.24}$$

Using this gauge, it is now possible to express the Lagrangian defined in Eq. 2.19 as

$$\begin{aligned} \mathcal{L} =& \frac{1}{2}(\partial_\mu h)(\partial^\mu h) + \frac{1}{4} g_W^2 (W_\mu^+ W^{\mu -})(v+h)^2 \\ &+ \frac{g_W^2 + g'^2}{8}(Z_\mu Z^\mu)(v+h)^2 \\ &+ \frac{\mu^2}{2}(v+h)^2 - \frac{\lambda}{16}(v+h)^4. \end{aligned} \tag{2.25}$$

Here the $W_\mu^\pm$ fields are those associated with the massive W bosons. The field for the massive neutral Z boson can be identified as a linear combination of the fields W_μ^3 and B_μ, defined in Eq. 2.16, as:

$$Z_\mu = \frac{g_W W_\mu^3 - g' B_\mu}{\sqrt{g_W^2 + g'^2}}, \tag{2.26}$$

and similarly the photon, with no associated mass term, can be identified as

$$A_\mu = \frac{g' W_\mu^3 + g_W B_\mu}{\sqrt{g_W^2 + g'^2}}. \tag{2.27}$$

The masses of the heavy bosons arising from this mechanisms can be written as

$$m_W = \frac{g_W v}{2}, \quad m_Z = \frac{v}{2}\sqrt{g_W^2 + g'^2} = \frac{m_W}{\cos\theta}. \tag{2.28}$$

Furthermore, the excitation of the field h can lead to the production of an additional massive boson, the *Higgs boson*, with a mass of

$$m_H = \sqrt{2|\mu|^2}. \tag{2.29}$$

In addition to providing masses to the heavy bosons in the SM, the Higgs mechanism also gives rise to fermion masses. It is possible to introduce terms into the Lagrangian, which describe the interactions of the fermions with the Higgs field, which are invariant under transformations in the $SU(2)_L \times U(1)_Y$ group. These terms are expressed as

$$\mathscr{L}_{\text{Yukawa}} = \sum_f -\frac{y_f v}{\sqrt{2}}\bar{\psi}_f \psi_f - \frac{y_f}{\sqrt{2}}\bar{\psi}_f \psi_f h, \tag{2.30}$$

where the fundamental parameters y_f are known as the *Yukawa couplings* of the fermions. These can be parametrised in terms of the observed particle masses as

$$y_f = \sqrt{2}\frac{m_f}{v}. \tag{2.31}$$

In the quark sector, an additional matrix needs to be introduced to account for the difference between the eigenstate basis for weak interactions and the mass-eigenstate basis. The transformation from the mass eigenstate basis is achieved using the *Cabbibo-Kobayashi-Maskawa (CKM) matrix* [3]:

$$V_{CKM} = \begin{pmatrix} V_{ud} & V_{cd} & V_{td} \\ V_{us} & V_{cs} & V_{ts} \\ V_{ub} & V_{cb} & V_{tb} \end{pmatrix} \tag{2.32}$$

The CKM matrix allows for transitions between different quark flavours using weak charged-current interactions. The parameters of the CKM matrix are measured experimentally. An almost diagonal matrix is observed, with small off-diagonal entries. This means that transitions between quarks of the same generation are preferred in charged-current interactions, e.g. $c \rightarrow s$, with $c \rightarrow d$ transitions also possible at a lower rate.

The Higgs mechanism completes the current formulation of the Standard Model of Particle Physics. It describes all currently observed particles and their interactions and explains the strong and electroweak interaction via particle properties such as colour charge, electric charge, isospin and hypercharge. The SM also describes a mechanism through which massive bosons and fermions acquire mass. This theory has several free parameters that are not predicted and require measurement. The precise measurement of these SM parameters is one of the main goals of particle physics today. The free parameters are:

- The 9 mass parameters of the up and down-type quarks and the charged leptons.
- The 3 coupling strengths of the strong interaction (g_s) and the electroweak interaction g_W and g'.
- The CKM matrix, which can be parametrised in terms of 3 mixing angles and 1 complex phase.

- The mass of the Higgs boson and the vacuum expectation value of the Higgs field.
- Additional free parameters related to CP violation in strong interactions and neutrino oscillations. These are not discussed in the context of the thesis.

2.3 Higgs Phenomenology at the LHC

The observation of the Higgs boson was one of the primary goals of the LHC since the start of operations. At a proton-proton collider, the Higgs boson can be produced in different ways and measured in various decay channels, which are discussed in the following. The coupling of the Higgs boson to other particles increases with mass, as described above, so the Higgs couples more strongly to massive particles, which is relevant for both the production and decay modes.

First, one must understand how particles are produced in hadronic collisions and define the cross-section. Protons are composite particles, and therefore in pp collisions, the hard scattering interactions occur between the *partons* (quarks or gluons) within the proton. This also means that the longitudinal momentum of the colliding partons is not known precisely. It is possible to define the partonic cross-section $\hat{\sigma}_{ab\to X}$ for two partons a and b, which carry fractions x_1 and x_2 of the proton momentum respectively. This cross-section describes the probability of the transition of the partons from their initial state to a final state X, and it can be computed from the interaction Lagrangian. It is possible to derive the inclusive cross-section for the specific process $pp \to X$, according to the factorisation theorem [19], as

$$\sigma_{pp\to X} = \sum_{a,b} \int_0^1 dx_1 \int_0^1 dx_2 f_a(x_1, \mu_F) f_b(x_2, \mu_F) \int d\sigma_{ab\to X}(x_1 P_1, x_2 P_2, \mu_R, \mu_F). \tag{2.33}$$

Here, the *parton distribution functions* (PDFs) $f_i(x_i, \mu_F)$ describe the probability of a parton i to participate in a hard scattering interaction with a momentum of $p_i = x_i P_i$, with P_i being the proton momentum. The PDFs depend on the energy scale at which the collisions occur but are independent of the results of the interaction. The energy scale is parametrised in terms of the *factorisation scale* μ_F, which can be interpreted as the scale that separates long and short-distance physics, i.e. the maximum value of transverse momentum where partons are still considered as part of the proton. An example of the process $ab \to X$ can be a specific production mode of the Higgs boson, which make up the inclusive cross-section $pp \to H$.

There are four main Higgs production modes, which are of relevance in the pp collisions at the LHC: *gluon-gluon fusion* (ggF), *vector boson fusion* (VBF), *associated production with a vector boson* (VH or *Higgsstrahlung*) and *associated production with top quarks* ($t\bar{t}H$). These four production modes are illustrated in Fig. 2.3. Their cross-section depends on the center-of-mass energy, and the cross-sections at the

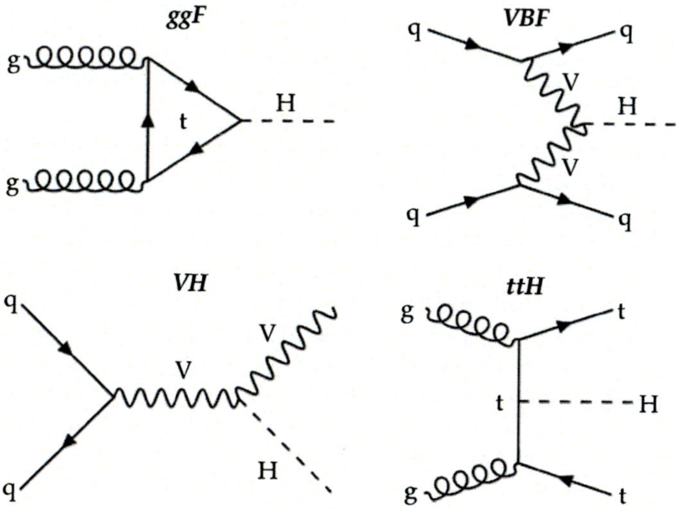

Fig. 2.3 Leading order (LO) Feynman diagrams for the main Higgs boson production modes at the LHC: gluon-gluon fusion (ggF), vector-boson fusion (VBF), associated production with a vector boson (VH) and associated production with top quarks (ttH)

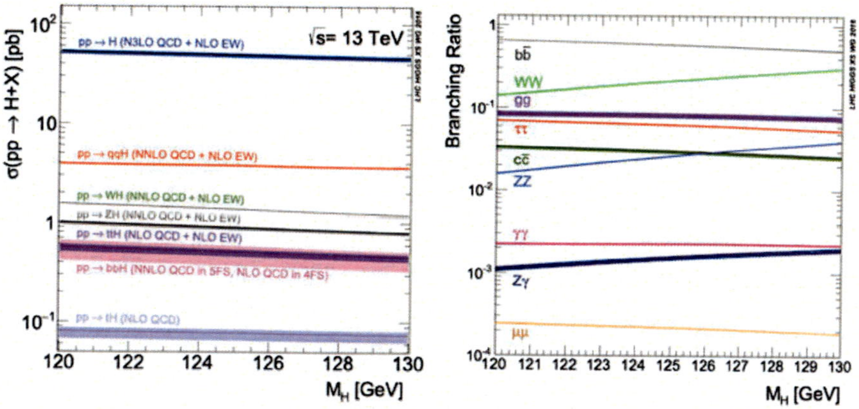

Fig. 2.4 Predicted SM Higgs production cross-sections at $\sqrt{s} = 13$ TeV (left) and SM Higgs boson decay branching ratios (right) as a function of Higgs boson mass. From [20]

LHC center-of-mass energy of $\sqrt{s} = 13$ TeV are illustrated in Fig. 2.4 (left). Other production modes are also possible, but have lower cross-sections.

The most common Higgs production mode at the LHC is *gluon-gluon fusion* (ggF), with a cross-section of $\sigma_{ggF} = 48.6 \pm 2.4$ pb [20]. The large production cross-section arises from the fact that gluons significantly contribute to the parton distribution function [21]. Gluons are massless, so the Higgs boson is produced via a top quark loop. Other kinds of loops are possible but are suppressed due to the lower masses of the other fermions. While this is the most abundant Higgs production mode, it is difficult to exploit it for the $H \rightarrow b\bar{b}$ and $H \rightarrow c\bar{c}$ decays discussed in this thesis due to the prevalence of non-resonant QCD processes in pp collisions.

The second most common production mode is *vector boson fusion* (VBF), with a cross-section of $\sigma_{VBF} = 3.77 \pm 0.09$ pb [20]. Here, two quarks scatter in an interaction mediated by two off-shell vector bosons (W or Z), which then can produce a Higgs boson. This production mode is easier to identify experimentally, as it is characterised by two forward jets in the event, in addition to the decay products of the Higgs boson [22].

In *VH production*, the Higgs boson is produced in association with a W or Z boson. This process has a small cross-section of $\sigma_{VH} = 2.24 \pm 0.14$ pb [20]. However, this production mode can provide a very clean signature in the detector when selecting events with a leptonic decay of the vector boson, significantly reducing the QCD background. Consequently, it is the most sensitive for searching for Higgs decays into quarks and is commonly used in the searches for Higgs decays into b- and c-quarks.

The final Higgs production mode discussed here is the *associated production with a top quark pair* ($t\bar{t}H$). With a cross-section of $\sigma_{ttH} = 0.51 \pm 0.04$ pb this process is rare [20]. However, it offers a possibility to probe the Higgs coupling to top quarks directly.

The lifetime of the Higgs boson is very short, with a predicted total width of the SM Higgs of around 4 MeV [23], which implies a lifetime of $O(10^{-22}$ s). Consequently, the Higgs can only be measured via its decay products. The Higgs boson coupling increases with mass, so decays into heavy particles are preferred. Additionally, decays of the Higgs boson via quark loops are possible, similarly to the mechanism in the ggF production mode.

Relative decay rates of particles are described by a *branching ratio* (BR), which represents the probability that a parent particle P decays into a final state X_i. The branching ratio is calculated as

$$\mathrm{BR}(P \rightarrow X_i) = \frac{\Gamma(P \rightarrow X_i)}{\Gamma_P} = \frac{\Gamma(P \rightarrow X_i)}{\sum_j \Gamma(P \rightarrow X_j)}, \tag{2.34}$$

where $\Gamma(P \rightarrow X_i)$ is the *partial decay width*, and Γ_P is the total width of the particle, which can be written as the sum of all possible decay modes. Figure 2.4 (right) shows the expected branching ratios of the Higgs boson as a function of Higgs mass. The most abundant decay mode is $H \rightarrow b\bar{b}$ with a branching ratio of 58.1%. The next most common decay into quarks is $H \rightarrow c\bar{c}$, with a branching ratio of 2.9%, which will be discussed in more detail in Chap. 4. However, decays into quarks are difficult to observe due to an overwhelming background from non-resonant QCD events.

A particle with properties consistent with the SM Higgs boson was discovered in 2012 by the ATLAS and CMS collaborations, using the $H \rightarrow \gamma\gamma$, $H \rightarrow ZZ \rightarrow 4\ell$ and $H \rightarrow WW^* \rightarrow 2\ell 2\nu$ channels [25, 26]. These channels have excellent sensitivity and provide clean signatures to identify the Higgs boson candidate. Since the discovery, many more of the Higgs boson production and decay modes have been

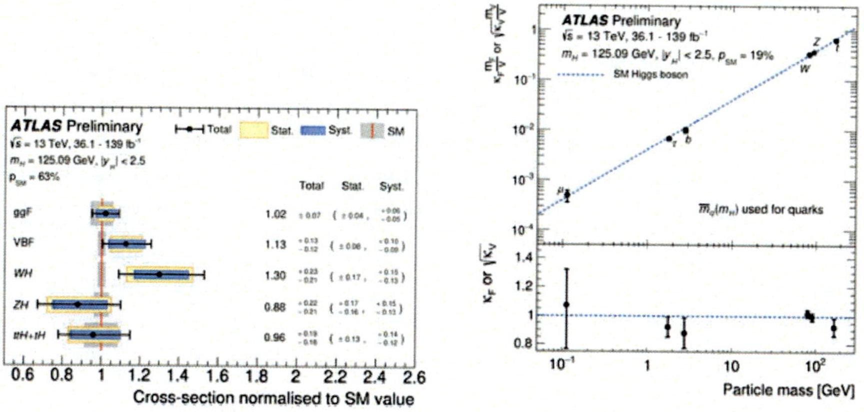

Fig. 2.5 Measured cross sections for ggF, VBF, WH, ZH and $t\bar{t}H$ production modes (left) and measured coupling-strength modifiers $\kappa_F \frac{m_F}{v}$ (for fermions) and $\sqrt{\kappa_V}\frac{m_V}{v}$ (for weak gauge bosons) (right). From [24]

measured, with summaries of the results shown in Fig. 2.5. The Higgs boson production cross-sections and the measured couplings to the different particles show good agreement with the SM expectations.

2.4 Theoretical Predictions for Hadron Colliders

In measurements of the Higgs boson, the collected collision data has to be compared to theoretical predictions and in pp collisions computing these theoretical predictions is very complex. It requires solving multi-dimensional integrals such as Eq. 2.33 and the calculation of QCD interactions in the hard scatter. An illustration of a pp collision is shown in Fig. 2.6. Due to their complex nature, the computations are performed using numerical methods known as Monte-Carlo (MC) techniques [28]. The simulation of pp collision events is done in several steps.

First, the *hard scattering* interaction is computed through the calculation of the hadronic cross-section $\sigma_{pp\to X}$, as illustrated in Eq. 2.33. This includes the calculation of the matrix element, i.e. $\sigma_{ab\to X}$, according to the interaction Lagrangian, using the PDFs and taking into account the hadronisation and factorisation scales [10, 11]. In the hard scatter, highly energetic particles are produced. In the case of gluons and quarks, these are initially too energetic to form bound states, and they produce additional quarks and gluons through radiation or gluon splitting. This step is simulated separately and known as *parton shower generation*. In this case, perturbative QCD can be used to perform the calculations. Once the energy of the gluons has reduced to O(1) GeV, the quarks and gluons form bound states through *hadronisation*. Perturbative QCD can no longer be used at these energy scales, and phenomenological

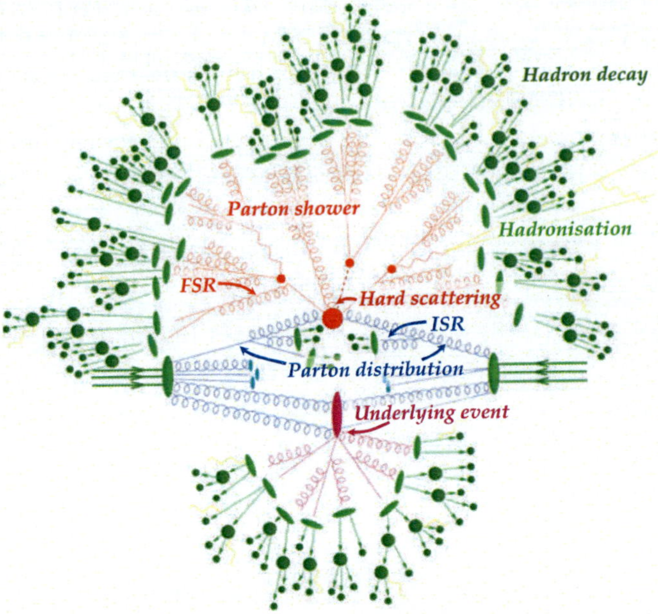

Fig. 2.6 Sketch of a proton-proton collision as simulated in a Monte-Carlo event generator. Modified from [27]

Table 2.1 Summary of Monte Carlo generators used in this thesis, which part of the *pp* interaction they are used to simulate, and the order at which the processes are simulated

MC generator	Hard scatter	Parton shower	Hadronisation	Underlying event	Accuracy
SHERPA [30–32]	×	×	×	×	NLO
MADGRAPH5+ AMC@NLO [33]	×			×	NLO
POWHEG [34, 35]	×			×	NLO
PYTHIA8 [36]		×	×	×	LO
HERWIG7 [37]		×	×	×	LO

QCD models are used instead to convert parton showers into outgoing hadrons. It is also possible to have *initial state radiation* (ISR), and *final state radiation* (FSR) in the form of gluons, or, less frequently, photons, radiated from the ingoing and outgoing partons, which have to be considered. Next to the hard scattering, the remaining partons present in the proton can also interact and these interactions are simulated as the *underlying event*. Finally, in the LHC, multiple protons collide simultaneously, which means several interactions can happen at the same time. This is known as *pile-up* [29] and is taken into account by adding soft QCD events to the simulated event before processing it further.

Several MC generators are available for this chain of computations. The different generators either simulate all steps described above or can be interfaced with one another to simulate the various stages. A summary of the MC generators used in this thesis is provided in Table 2.1.

Once the events are generated, they need to pass through a simulation of the ATLAS detector to account for and match any reconstruction performed on real collision data. The ATLAS simulation chain contains four steps [38] and is implemented within the software framework ATHENA [39]. First, events are generated using one of the available MC generators. In particular, the matrix element for the hard scatter process is calculated, as well as the parton shower and hadronisation. The truth information for each stable simulated particle is stored and then processed through the detector simulation. The *detector simulation* is the most computationally intensive step of the simulation chain. The full detector geometry, material and response are simulated using GEANT4 [40]. Particles are propagated through the detector, and the interactions with the sensitive detector material are recorded as *hits*. The energy deposit is stored for each interaction, along with the position and timing information. The hits are transformed into digital signals (*digitisation*), simulating the output provided by the different parts of the detector. The information is stored as a *Raw Data Object* (RDO). Finally, particles are constructed from the detector information in the *reconstruction* step. Particular reconstruction algorithms are used for different physics objects, as described in Sect. 3.3. The output is stored in the Analysis Object Data (AOD) format, which can be further processed in physics analyses to perform comparisons to data and measurements.

References

1. Thomson M (2013) Modern particle physics. Cambridge University Press, New York. ISBN: 978-1-107-03426-6
2. Perkins DH (1982) Introduction to high energy physics. ISBN: 978-0-521-62196-0
3. Tanabashi M et al (2018) Review of particle physics. Phys Rev D 98:030001. https://doi.org/10.1103/PhysRevD.98.030001
4. Ellis J (2014) The discovery of the gluon. Int J Mod Phys A 29(31):1430072. ISSN: 1793-656X. https://doi.org/10.1142/s0217751x14300725
5. Di Lella L, Rubbia C (2015) The discovery of the wand Z particles. Adv Ser Direct High Energy Phys 23:137–163. https://doi.org/10.1142/9789814644150_0006
6. Noether E (1918) Invariante Variationsprobleme. ger. In: Nachrichten von der Gesellschaft der Wissenschaften zu Göttingen,Mathematisch-Physikalische Klasse 1918:235–257. http://eudml.org/doc/59024
7. Gross DJ (2005) The discovery of asymptotic freedom and the emergence of QCD. Proc Nat Acad Sci 102:9099–9108. https://doi.org/10.1073/pnas.0503831102
8. Gell-Mann M (1964) A schematic model of Baryons and mesons. Phys Lett 8:214–215. https://doi.org/10.1016/S0031-9163(64)92001-3
9. Zweig G (1964) An SU(3) model for strong interaction symmetry and its breaking. Version 2. In: Developments in the quark theory of hadrons, vol 1, 1964 - 1978. Lichtenberg DB, Peter Rosen S (eds) pp 22–101
10. Mangano ML (1999) Introduction to QCD. https://doi.org/10.5170/CERN-1999-004.53

11. Collins JC, Soper DE, Sterman G (2004) Factorization of hard processes in QCD. arXiv:hep-ph/0409313 [hep-ph]
12. Glashow SL (1961) Partial-symmetries of weak interactions. Nucl Phys 22(4):579–588. ISSN: 0029-5582. https://doi.org/10.1016/0029-5582(61)90469-2
13. Salam A Weak and electromagnetic interactions. Selected papers of abdus salam, pp 244–254. https://doi.org/10.1142/9789812795915_0034
14. Weinberg S (1967) A model of leptons. Phys Rev Lett 19:1264–1266. https://doi.org/10.1103/PhysRevLett.19.1264
15. Englert F, Brout R (1964) Broken symmetry and the mass of Gauge vector mesons. Phys Rev Lett 13:321–323. https://doi.org/10.1103/PhysRevLett.13.321
16. Higgs PW (1964) Broken symmetries, massless particles and gauge fields. Phys Lett 12(2):132–133. ISSN: 0031-9163. https://doi.org/10.1016/0031-9163(64)91136-9
17. Higgs PW (1964) Broken symmetries and the masses of Gauge Bosons. Phys Rev Lett 13:508–509. https://doi.org/10.1103/PhysRevLett.13.508
18. Higgs PW (1966) Spontaneous symmetry breakdown without massless Bosons. Phys Rev 145:1156–1163. https://doi.org/10.1103/PhysRev.145.1156
19. Collins JC, Soper DE, Sterman GF (1989) Factorization of hard processes in QCD. Adv Ser Direct High Energy Phys 5:1–91 arXiv:hep-ph/0409313
20. de Florian D et al (2016) Handbook of LHC Higgs cross sections: 4. Deciphering the nature of the Higgs sector. CERN yellow reports: monographs. Geneva: CERN. https://doi.org/10.23731/CYRM-2017-002
21. Georgi HM et al (1978) Higgs Bosons from two-gluon annihilation in proton-proton collisions. Phys Rev Lett 40:692–694. https://doi.org/10.1103/PhysRevLett.40.692
22. Bolzoni P et al (2012) Vector boson fusion at next-to-next-to-leading order in QCD: standard model Higgs boson and beyond. Phys Rev D 85:035002. https://doi.org/10.1103/PhysRevD.85.035002
23. Dittmaier S et al (2011) Handbook of LHC Higgs cross sections: 1. Inclusive observables. CERN yellow reports: monographs. Geneva: CERN. https://doi.org/10.5170/CERN-2011-002
24. ATLAS Collaboration (2021) Combined measurements of Higgs boson production and decay using up to 139 fb-1 of proton-proton collision data at ps = 13 TeV collected with the ATLAS experiment. Technical report Geneva: CERN. http://cds.cern.ch/record/2789544
25. ATLAS Collaboration (2012) Observation of a new particle in the search for the standard model Higgs boson with the ATLAS detector at the LHC. Phys Lett B 716:1. https://doi.org/10.1016/j.physletb.2012.08.020. arXiv:1207.7214 [hep-ex]
26. CMS Collaboration (2012) Observation of a new boson at a mass of 125GeV with the CMS experiment at the LHC. Phys Lett B 716:30. https://doi.org/10.1016/j.physletb.2012.08.021. arXiv:1207.7235 [hep-ex]
27. Höche S (2015) Introduction to parton-shower event generators. In: Theoretical advanced study institute in elementary particle physics: journeys through the precision frontier: amplitudes for colliders, pp 235–295. https://doi.org/10.1142/9789814678766_0005. arXiv:1411.4085 [hep-ph]
28. Cattani C, Pop F (2014) High performance numerical computing for high energy physics: a new challenge for big data science. Adv High Energy Phys 2014:507690. https://doi.org/10.1155/2014/507690
29. Marshall Z (2014) Simulation of pile-up in the ATLAS experiment. J Phys Conf Ser 513(2):022024. https://doi.org/10.1088/1742-6596/513/2/022024
30. Gleisberg T et al (2009) Event generation with SHERPA 1.1. JHEP 02:007. https://doi.org/10.1088/1126-6708/2009/02/007. arXiv:0811.4622 [hep-ph]
31. Höche S et al (2009) QCD matrix elements and truncated showers. JHEP 05:053. https://doi.org/10.1088/1126-6708/2009/05/053. arXiv:0903.1219 [hep-ph]
32. Cascioli F, Maierhöfer P, Pozzorini S (2012) Scattering amplitudes with open loops. Phys Rev Lett 108:111601. https://doi.org/10.1103/PhysRevLett.108.111601. arXiv:1111.5206 [hep-ph]

33. J. Alwall et al. "The automated computation of tree-level and next-to-leading order differential cross sections, and their matching to parton shower simulations". In: JHEP 07 (2014), p. 079. DOI: https://doi.org/10.1007/JHEP07(2014)079.arXiv:1405.0301 [hep-ph]
34. Frixione S, Nason P, Oleari C (2007) Matching NLO QCD computations with parton shower simulations: the POWHEG method. JHEP 11:070. https://doi.org/10.1088/1126-6708/2007/11/070. arXiv:0709.2092 [hep-ph]
35. Alioli S et al (2010) A general framework for implementing NLO calculations in shower Monte Carlo programs: the POWHEGBOX. JHEP 06:043. https://doi.org/10.1007/JHEP06(2010)043. arXiv:1002.2581 [hep-ph]
36. Sjöstrand T et al (2015) An introduction to PYTHIA 8.2. Comput Phys Commun 191:159. https://doi.org/10.1016/j.cpc.2015.01.024. arXiv:1410.3012 [hep-ph]
37. Bellm J et al (2016) Herwig 7.0/Herwig++ 3.0 release note. Eur Phys J C 76(4):196. https://doi.org/10.1140/epjc/s10052-016-4018-8. arXiv:1512.01178 [hep-ph]
38. Aad G et al (2010) The ATLAS simulation infrastructure. Eur Phys J C 70(3):823–874. ISSN: 1434-6052. https://doi.org/10.1140/epjc/s10052-010-1429-9
39. Aimar A, Harvey J, Knoors N (2005) Computing in high energy physics and nuclear physics 2004. Geneva: CERN. https://doi.org/10.5170/CERN-2005-002
40. Agostinelli S et al (2002) GEANT4–a simulation toolkikt. GEANT4. A simulation toolkit. Nucl Instrum Methods Phys Res, A 506:250–303. 54 p. https://doi.org/10.1016/S0168-9002(03)01368-8

Chapter 3
Experimental Particle Physics with the ATLAS Detector

3.1 The Large Hadron Collider

The Large Hadron Collider (LHC) [1] at CERN is the world's largest particle accelerator, with a circumference of 27 km. It accelerates protons, or heavy ions, to high energies and collides them to allow the study of particles produced in this collision and their decay products in complex detector systems. In this thesis, only data from proton-proton collisions were studied; therefore, only *pp* collisions are discussed in the following. Before protons can enter the LHC, they need to pass through a chain of pre-accelerators, as shown in Fig. 3.1. The protons are obtained from a bottle of hydrogen gas, which is ionised using an electric field. A linear accelerator (Linac 2, replaced by Linac 4 in 2020) then accelerates the protons to energies of 50 MeV. This beam is then injected into the Proton Synchrotron Booster (PSB), which consists of four stacked synchrotron rings. The PSB accelerates the protons to 1.4 GeV, shapes the beam into batches, also called bunches, and then injects it into the Proton Synchrotron (PS). The PS increases the beam energy to 25 GeV, followed by the Super Proton Synchrotron (SPS), which accelerates the protons to 450 GeV. Finally, the beams are injected into the LHC, where the protons circulate in two beam pipes in opposite directions. Here, they are accelerated using radio-frequency cavities, kept on a circular path using superconducting dipole magnets with a magnetic field of 8.33 T and shaped or focused using higher-order magnets. This process continues until the bunches reach an energy of 6.5 TeV per beam, spaced by 25 ns in time. The two beams are then brought to collision at four interaction points, which are the locations of the main LHC experiments: ALICE [2], ATLAS [3], CMS [4] and LHCb [5].

The ALICE experiment is dedicated to studying heavy-ion collisions, particularly the resulting quark-gluon-plasma, which gives rise to *collective* phenomena and can behave as a perfect fluid. The heavy ion collisions result in events with many decay products, including heavy flavour quarks, which are a sensitive probe of the quark-gluon plasma. The LHCb experiment predominantly focuses on studying

M. Mironova, *Search for Higgs Boson Decays to Charm Quarks with the ATLAS Experiment and Development of Novel Silicon Pixel Detectors*, Springer Theses,
https://doi.org/10.1007/978-3-031-36220-0_3

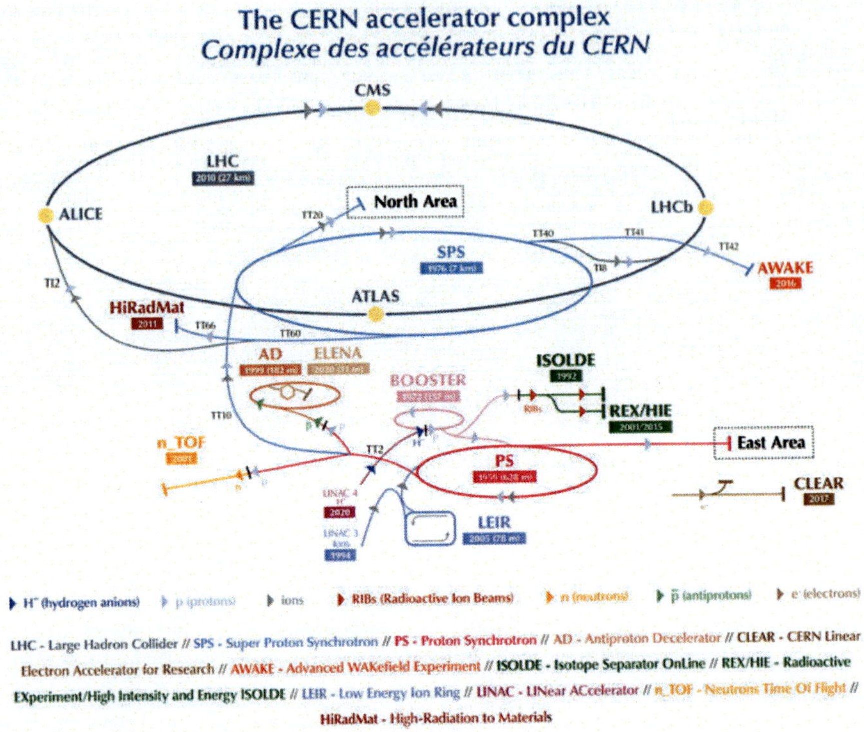

Fig. 3.1 Sketch of the LHC accelerator complex. From [6]

heavy flavour physics. It has an asymmetric detector, as B-mesons are primarily produced in the very forward direction. The ATLAS and CMS experiments are large general-purpose detectors, and the ATLAS detector is further described in Sect. 3.2.

3.1.1 Luminosity, Center-of-Mass Energy and Pile-up

One quantity of particular importance in collider physics is the instantaneous luminosity $\mathcal{L}_{\text{inst}}$, which relates the rate at which a particular physics process occurs to its cross-section:

$$\frac{dN}{dt} = \mathcal{L}_{\text{inst}} \times \sigma \tag{3.1}$$

The instantaneous luminosity can be calculated from the collider parameters as follows:

$$\mathcal{L}_{\text{inst}} = \frac{N_1 N_2 f N_b}{4\pi \sigma_x \sigma_y} \tag{3.2}$$

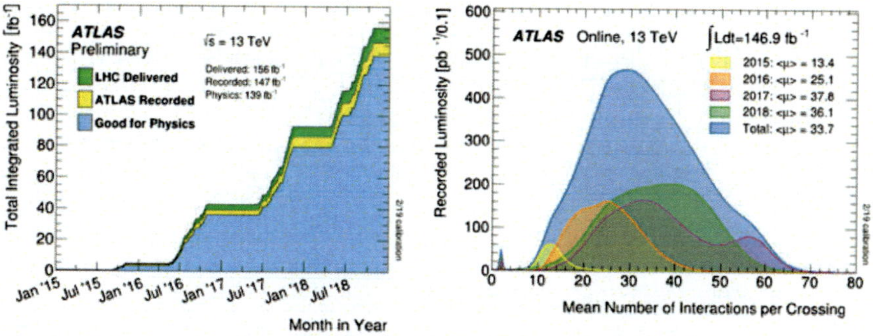

Fig. 3.2 Integrated delivered and recorded luminosity as a function of time in the ATLAS experiment (left). Luminosity-weighted distribution of the mean number of interactions per bunch crossing for the full Run 2 pp collision dataset (right). From [8]

where N_1 and N_2 are the number of protons in each of the two colliding bunches of particles, f is the revolution frequency of the collider, N_b is the number of bunches, and σ_x and σ_y are the widths of the beam density distributions in the x- and y-direction, respectively. For the first run of the LHC the maximum instantaneous luminosity was $\mathscr{L} = 0.8 \times 10^{34}\ \text{cm}^{-2}\text{s}^{-1}$ [7]. For the second run of the LHC, the luminosity was even higher with $\mathscr{L} = 2.1 \times 10^{34}\ \text{cm}^{-2}\text{s}^{-1}$ [8].

The integrated luminosity $\mathscr{L}_{\text{int}} = \int \mathscr{L} dt$ can be used to calculate the total number of events produced for a process with a particular cross-section:

$$N = \mathscr{L}_{\text{int}} \times \sigma \tag{3.3}$$

Figure 3.2 (left) shows the integrated luminosity delivered by the LHC, and collected by the ATLAS experiment between 2015 and 2018 (Run 2). The total dataset corresponds to an integrated luminosity of $\mathscr{L} = (139.0 \pm 2.4)\ \text{fb}^{-1}$ [9].

A second important quantity to the LHC operation is the center-of-mass energy $\sqrt{s}$, denoted by the Mandelstam variable s. It corresponds to the maximum available energy for the production of particles. In the first run of the LHC, pp collision data was collected at $\sqrt{s}$ = 7 TeV and 8 TeV. For Run 2, the superconducting magnets were repaired, and a center-of-mass energy of 13 TeV was achieved.

In the ATLAS experiment, head-on collisions of two protons in a bunch-crossing, which produce hard scattering events, are of interest. However, each hard scatter event has additional soft pp interactions happening at the same time, known as *pile-up*, which can deteriorate the reconstruction quality of the main hard scattering event. *In-time pile-up* are additional soft interactions that occur in the same bunch-crossing, and *out-of-time pile-up* stems from neighbouring bunch-crossings. The effect of pile-up is characterised by the average number of interactions per event μ, defined as

$$\mu = \frac{\mathscr{L}_{\text{bunch}} \sigma_{inel}}{N_b f}, \tag{3.4}$$

where $\mathscr{L}_{\text{bunch}}$ is the instantaneous luminosity per bunch crossing, σ_{inel} is the inelastic scattering cross-section for pp collisions, N_b is the number of bunches and f is the collision frequency. The mean number of interactions per bunch crossing corresponds to the mean of the Poisson distribution of the number of interactions per crossing calculated for each bunch. This distribution is shown in Fig. 3.2 (right) for the various data-taking periods.

3.2 The ATLAS Detector

The ATLAS detector [3] consists of cylindrical detector layers and endcaps, as shown in Fig. 3.3. This layout is chosen due to the general-purpose philosophy of the detector. The ATLAS experiment was designed to observe a broad range of physical phenomena; hence, the detector covers almost the whole solid angle. Due to the 25 ns bunch structure of the LHC and the large datasets collected, the different detector components require fast, radiation-hard sensors and readout electronics. In addition, a good spatial resolution is required to distinguish multiple events occurring simultaneously. The ATLAS detector has a total length of 44 m, a height of 25 m and a weight of 7000 t [3].

In the coordinate system used for the ATLAS detector, the z-axis is defined to be the direction of the beam. The x-axis points towards the centre of the LHC ring, and the y-axis points upwards from the interaction point. Using polar coordinates, the azimuthal angle ϕ is defined in the xy-plane and the angle θ is measured upwards

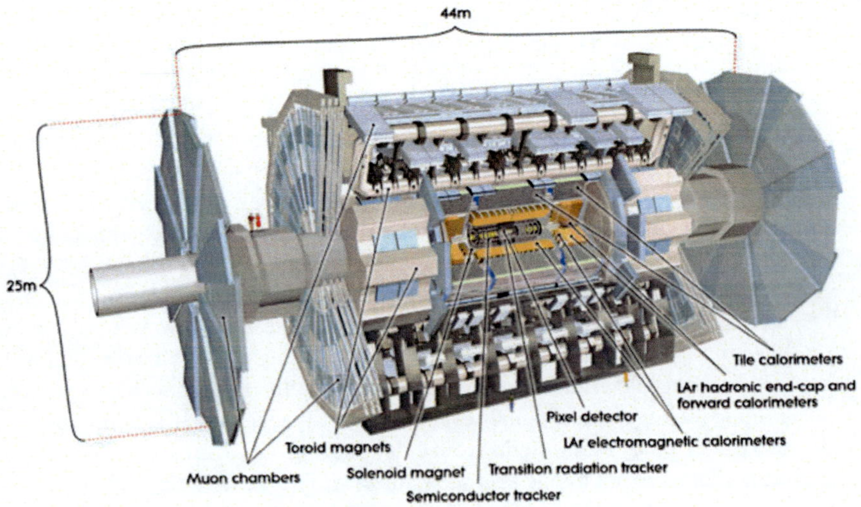

Fig. 3.3 Cut-away view of the ATLAS detector. From [3]

from the beam axis. One crucial quantity in particle physics is the rapidity y, which is a measure of the direction of a particle and is defined as:

$$y = \frac{1}{2} \ln \left(\frac{E + p_L}{E - p_L} \right),$$

where E is the energy and p_L is longitudinal momentum. For ultra-relativistic particles with $p >> m$ the pseudo-rapidity $\eta = -\ln\left(\tan\left(\frac{\theta}{2}\right)\right)$ is an appropriate approximation of the rapidity. The pseudo-rapidity only depends on the polar angle θ and is useful as the intervals of pseudo-rapidity, $\Delta\eta$, are Lorentz invariant under boosts along the longitudinal axis. Another important kinematic variable is ΔR, which defines the angular separation between two objects in the detector as a function of η and ϕ:

$$\Delta R_{1,2} = \sqrt{\Delta\phi_{12}^2 + \Delta\eta_{12}^2} = \sqrt{(\phi_1 - \phi_2)^2 + (\eta_1 - \eta_2)^2}.$$

Finally, when considering energy and momenta, the transverse components of these quantities are commonly used, since the momentum of the colliding partons along the z-axis is not known. Thus, the quantities p_T and E_T are defined as $p_T = p \sin\theta$ and $E_T = E \sin\theta$.

3.2.1 Magnet System

One of the critical components of the ATLAS detector is its magnet system. Magnetic fields are crucial for measuring the momentum of charged particles, as charged particles follow curved trajectories in a magnetic field due to the Lorentz force. The radius of the particle trajectory R is given by $R = \frac{p\cos\lambda}{0.3B}$, where p is the momentum of the particle (in units of GeV/c), λ is the angle of the particle with respect to the z-axis, B is the magnetic field (in T), and R is the radius (in m). Thus, high magnetic fields are needed to increase the curvature of high-p_T particles and measure their momentum precisely.

The ATLAS experiment has a complex magnet system with four superconducting solenoids, which shape the magnetic field within the inner detector and the muon spectrometer [10]. The central solenoid provides a 2 T magnetic field along the z-axis, such that particles are deflected along the ϕ-direction. Additionally, the barrel and endcap toroids in the muon detector yield a magnetic field between 0.5 and 1 T and are used to deflect muons in the η-direction to measure their momentum precisely.

3.2.2 Inner Detector

The most central part of the ATLAS detector is the Inner Detector (ID) [12, 13]. Its purpose is to track charged particles from the interaction region until they reach the calorimeter and measure their momentum by determining the curvature of the tracks in the magnetic field. The ID covers a region of $|\eta| < 2.5$ and a radius between 3 cm and 1 m. The ID environment is characterised by a high track density, requiring a fast and radiation tolerant detector, as well as a high granularity for an excellent resolution of the reconstructed momenta and vertices. The ID consists of three sub-systems: the *Pixel Detector*, the *SemiConductor Tracker* and the *Transition Radiation Tracker*. An illustration of the ID is shown in Fig. 3.4.

Pixel Detector The silicon pixel detector is the innermost system [14]. It consists of three barrel layers and three endcap layers, is made of 1744 pixel modules with a total of 80 million pixels. The modules have a pixel pitch of 50×400 mm^2 and a sensor thickness of 250 mm. During the first long shutdown of the LHC (LS1), an additional pixel detector layer was installed: the Insertable B-Layer (IBL) [15]. This

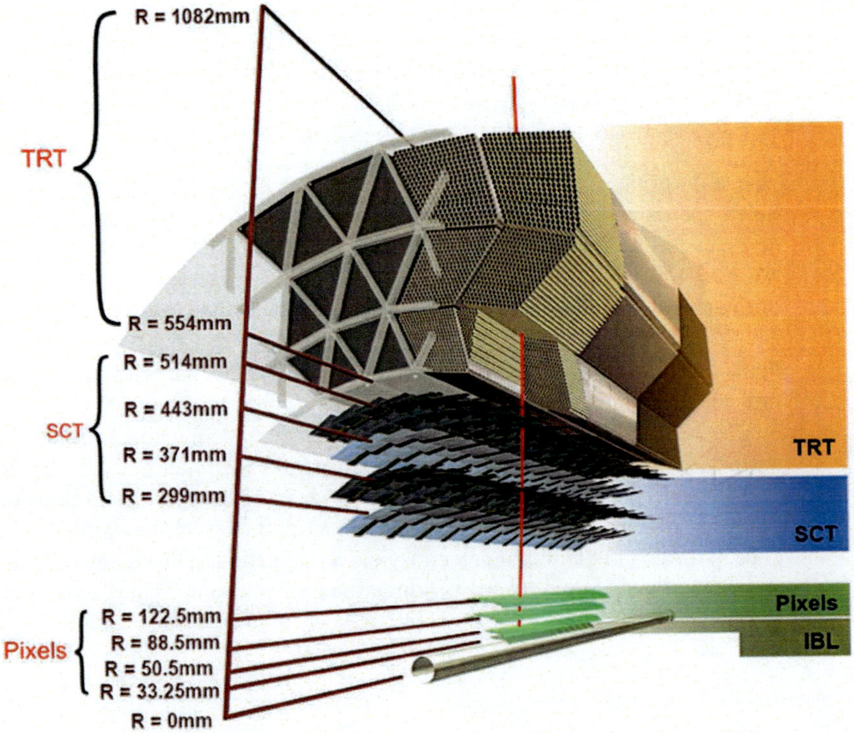

Fig. 3.4 Sketch of the ATLAS ID showing all its components, including the new insertable B-layer (IBL). The distances to the interaction point are also shown. From [11]

layer is located 33 mm away from the beamline and includes 12 million pixels with a pixel pitch of 50 × 250 mm. The IBL was introduced to significantly improve the tracking robustness and precision, as well as the flavour tagging performance. The current pixel detector provides a spacepoint resolution of 10 mm in the $r - \phi$ plane and 67 mm in the z-direction [16, 17].

SemiConductor Tracker (SCT) The next layer of the ID is the SCT, which consists of silicon microstrip modules that are arranged into four concentric barrel layers and two endcaps with nine disks each [18]. The pitch of the strips is 80 mm, the thickness of the detectors is 285 mm. Two strip detectors are combined in a module with a stereo angle of 40 mrad to provide hit reconstruction in two dimensions. The SCT modules have a resolution of 17 mm in the $r - \phi$ plane and 580 mm in the z-direction [19].

Transition Radiation Tracker (TRT) The final component of the ID is the TRT [20]. The TRT is a straw-tube tracker which consists of 300 thousand cylindrical drift tubes (straws) with a diameter of 4 mm. Each straw has a gold-plated anode wire and is filled with a gas mixture based on Xenon. When a charged particle traverses the tube, it ionises the gas. The resulting electrons drift towards the anode wire and cascade in the electric field. This produces a detectable signal which can be compared against two adjustable thresholds. The TRT contributes not only to the momentum measurement for charged particles, but also to particle identification. The space between the straws is filled with polyethylene, and when a relativistic particle traverses the boundary between the two materials, it can emit radiation (*transition radiation*), dependent on the Lorentz boost $\gamma \sim E/m$. This means that the amount of radiation produced can be used to determine the particle mass for particles of the same momentum. In the ATLAS detector, this helps distinguish between electrons and pions with momenta of 1 to 100 GeV. The drift tubes measure the transition radiation photons via the photoelectrons they generate, identified as characteristic large energy deposits. The TRT provides a position resolution of 130 mm in the $r - \phi$ ($z - \phi$) plane in the barrel (endcap).

Combining all three components of the ID, the track of a charged particle in the detector can be reconstructed, and from the curvature of the track its momentum can be determined. The momentum resolution σ_{p_T} of the ATLAS ID is

$$\frac{\sigma_{p_T}}{p_T} = 0.05\% \ p_T \ [\text{GeV}] \oplus 1\%,$$

where p_T is the particle momentum, and $\oplus$ denotes a sum in quadrature [3].

3.2.3 Calorimeter

The ATLAS detector uses calorimeters to measure the energy of neutral and charged particles, as well as missing transverse momentum [22]. The ATLAS calorimeters

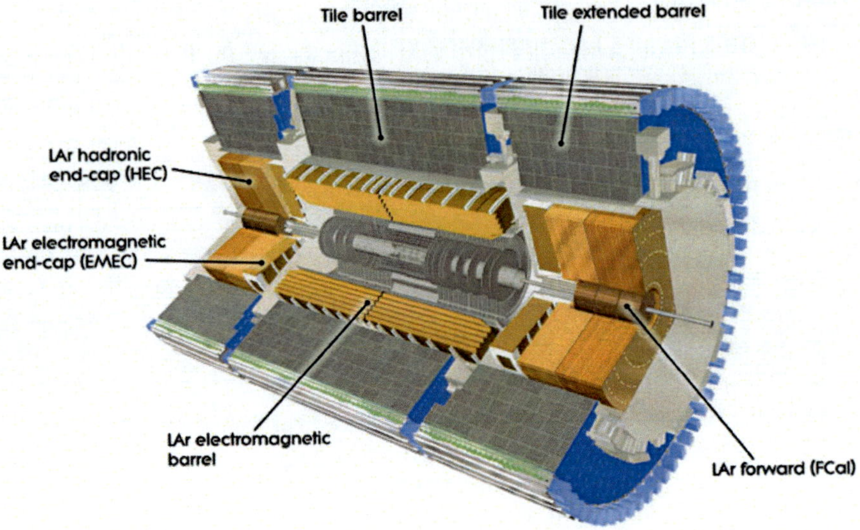

Fig. 3.5 Cut-away view of the ATLAS Calorimeter system. From [21]

are hermetic in ϕ and cover a range of $|\eta| < 4.9$. All of the calorimeters in ATLAS are *sampling* calorimeters, which means that they consist of multiple layers of active and passive material. The passive material is usually a dense material with a high atomic number, which induces cascades of particles (*showers*). The active layers measure the energy deposited by the showers, such as ionisation or scintillation light. In ATLAS, Liquid Argon (LAr) is mainly used as the active material [23].

The layout of the ATLAS calorimeters is shown in Fig. 3.5. The *electromagnetic* calorimeter (ECAL) is used to measure the energy deposited by electrons, photons and some of the energy carried by jets. It uses lead as the passive material, and liquid argon as the active material [24]. Electromagnetic showers occur via the electromagnetic interaction and are parametrised by the *radiation length* X_0, which is the average distance over which the energy of a high-energy electron will be reduced to $1/e$ of its original value. The electromagnetic calorimeters in ATLAS have a depth of at least 22 X_0.

The *hadronic* calorimeters are used to measure the energy of hadronic showers. The hadronic endcap and forward calorimeters both use LAr as the active material, and copper plates are used as the passive material in the endcap [25]. The forward calorimeter uses both copper and tungsten as the passive material [25, 26]. The tile calorimeter in the barrel uses steel as the absorbing material, and scintillating plastic tiles as the active material [27]. Hadronic showers in the detector are characterised by the nuclear interaction length λ, **defined as the mean distance travelled by a hadron** before interacting with a nucleus. The depth of the ATLAS hadronic calorimeters is $\sim$10λ. Thus, both the electromagnetic and hadronic calorimeters provide good containment of the particle showers.

The energy resolution of a calorimeter is described by

$$\frac{\sigma_E}{E} = \frac{a}{\sqrt{E}} \oplus \frac{b}{E} \oplus c.$$

The first term $a/\sqrt{E}$ is known as the *stochastic term*, which describes fluctuations related to Poisson statistics, for example shower or sampling fluctuations. The second term b/E is the *noise term*, which accounts for electronic noise in the detector readout. Finally, the *constant term c* describes instrumental effects, for example inefficiencies in the signal collection. For an electromagnetic calorimeter a typical energy resolution is $\frac{\sigma_E}{E} \approx \frac{10\%}{\sqrt{E}} + 0.7\%$ [28]. For a hadronic calorimeter the typical energy resolution is worse, due to the complexity of hadronic showers, with $\frac{\sigma_E}{E} \approx \frac{50\%}{\sqrt{E}} + 3\%$ [28].

3.2.4 *Muon Spectrometer*

The final and outermost detector of the ATLAS experiment is the muon spectrometer (MS) [30, 31]. Its purpose is to measure the momentum of muons, which are most of the charged particles traversing the detector after the hadronic calorimeter. The toroidal magnet system is critical for the muon momentum measurement. The precision of the measured muon momentum varies between 3% for low p_T muons ($2 < p_T < 250$ GeV) and 10% for high p_T muons. The muon system consists of different kinds of detectors to perform precision tracking and provide fast triggering information (Fig. 3.6).

Precision Tracking Chambers Precision tracking is achieved by a Monitored Drift Tube (MDT) detector system [32], which covers both the barrel and endcap region up to $\eta < 2.7$ in several layers. The MDTs provide a spatial resolution of 35 mm. In addition, the high occupancy region $2.0 < \eta < 2.7$ is covered by Cathode Strip Chambers (CSC) [33], which are multi-wire proportional chambers, with the cathodes segmented into strips. The CSCs can achieve a spatial resolution of 60 mm in the r direction and 5 mm in the ϕ direction.

Fast Trigger Chambers The muon trigger system is built using two different detectors. In the barrel region, three layers of Resistive Plate Chambers (RPCs) [34], are used. The endcap consists of 3–4 layers of Thin Gap Chambers [35]. The combination of the two systems provides a muon-based hardware trigger based on the coincidence of hits in different layers, as well as an additional measurement of the muon trajectory.

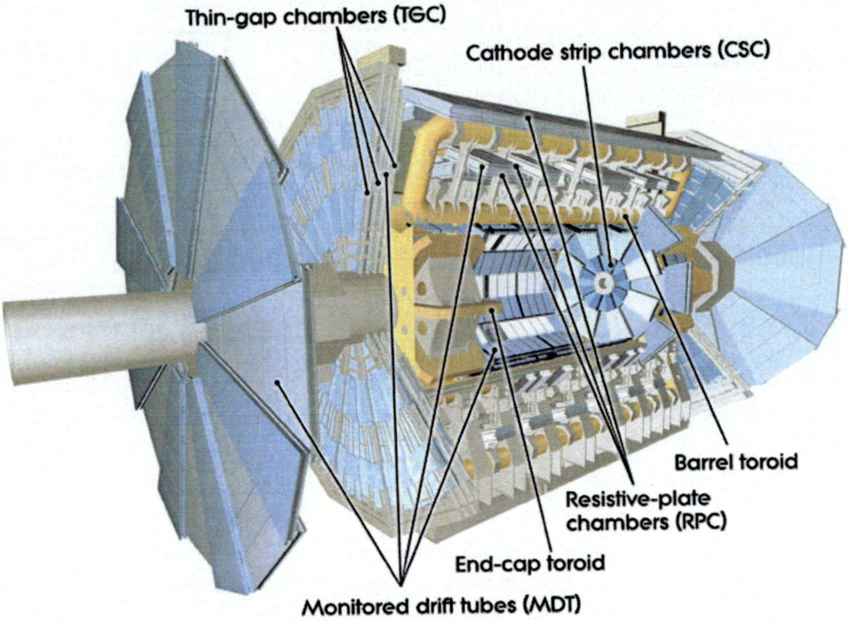

Fig. 3.6 Cut-away view of the ATLAS Muon system. From [29]

3.2.5 Trigger and DAQ

During Run 2 of the LHC, bunch-crossings occurred every 25 ns, i.e. at a rate of 40 MHz. Due to the limited available computing resources, only a subset of these events can be stored. Thus, it is crucial to quickly reject events without hard scattering interactions, while keeping relevant physics processes. This is achieved with the ATLAS trigger system [36], which reduces the data rate to 1 kHz. It consists of a hardware-based Level-1 (L1) trigger and a software-based high-level trigger (HLT).

Level 1 Trigger The L1 trigger uses coarse information from the calorimeters and muon spectrometer to identify high momentum objects, either large deposits in the calorimeter or high-p_T muons. The L1 trigger reduces the data rate to 100 kHz. It selects $\eta - \phi$ regions where interesting features have been found, and these Regions of Interest (RoI) are then passed to the HLT.

High Level Trigger The HLT integrates the RoI data with the complete detector information from all detector sub-systems and then performs a full offline-like event reconstruction. Then, trigger algorithms are used to select or reject the event. In the HLT, it is possible to identify more complex signatures, for example leptons, $E_{\mathrm{T}}^{\mathrm{miss}}$, events with b-jets or other physics objects.

A trigger menu defines the selection criteria for L1 and the HLT. Different triggers are used depending on the purpose of data-taking, for example for physics measurements, performance measurements and detector calibration.

3.3 Physics Object Reconstruction

Proton-proton collisions take place in the centre of the ATLAS detector, and the produced particles travel through the detector. They leave different signatures in the sub-detectors, as illustrated in Fig. 3.7, e.g. hits in the different layers of the inner detector and muon spectrometer, or energy deposits in the calorimeter. The signals from the detectors are processed and reconstructed as so-called physics objects, which are matched to final-state particles or observables in ATLAS, e.g. electrons, muons or jets. The reconstruction occurs in several steps. First, detector-level objects are reconstructed, such as *tracks*, *vertices* or *calorimeter clusters*. Then, they are combined into high-level objects, which correspond to the different particles. These physics objects are *reconstructed* and additional *identification* and *isolation* criteria are applied. Finally, the physics objects are *calibrated* to account for any detector effects which are different in data and simulation. This section provides an overview of the detector-level and high-level objects and the methods used to reconstruct them.

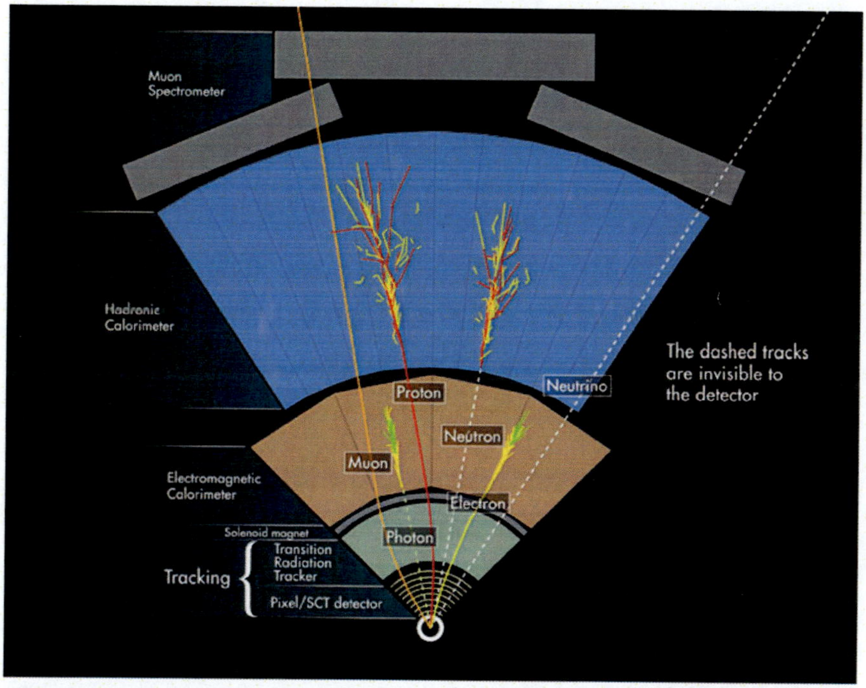

Fig. 3.7 View of the particle paths and experimental signatures in the transverse plane of the ATLAS detector. From [37]

3.3.1 Tracks, Vertices and Calorimeter Clusters

This section describes the *detector-level* objects reconstructed within the ATLAS detector, which are then used to construct the high-level physics objects.

Tracks A charged particle on a curved trajectory in the magnetic field of the ATLAS detector will interact with the sensitive material of the detector and leave energy deposits, which are registered as a signal by the detector. The tracks of particles can be reconstructed, starting by identifying clusters of energy deposits as *hits* [38]. Then, the track reconstruction algorithm begins, utilising the *inside-out* method [38], and starting from groups of three hits in the Pixel detector or SCT. Additional hits are then added to the track by using a Kalman filter [39] to identify the most likely additional hit based on the preliminary trajectory. With this method, hits can get assigned to multiple tracks. To resolve this ambiguity, tracks are ranked based on the quality of the track, utilising information about the number of hits, the origin of hits (Pixel or SCT) and the number of holes (layers with no hits on the predicted track). Additionally, a χ^2-fit is performed to assess the quality of the track, and high p_T tracks are preferred. Highly ranked tracks are then extended to the TRT using an *outside-in* approach. Finally, all tracks need to pass additional criteria, such as a minimum of 7 hits in the pixel detector and SCT, a maximum of one hole and a maximum of two shared hits. Kinematic criteria are also applied, requiring $p_T >$ 500 MeV and $|\eta| < 2.5$.

Vertices Identifying vertices, i.e. locations where interactions have occurred, is crucial to physics analyses. Of particular importance is the reconstruction of the primary vertex, which is where the *pp* hard scattering event has occurred [40, 41]. Due to the large pile-up in the LHC collisions, events commonly have multiple interaction vertices. The identification of vertices occurs in an iterative process. First, a starting point for the location of the vertex (*seed*) is identified by using the information of all good-quality reconstructed tracks. Then, all tracks that are incompatible with the computed vertex are removed, and the calculation of the vertex position is repeated. This process continues until the reconstructed vertex passes specific quality requirements. The previously discarded tracks are then used to find additional vertices. The *primary vertex* is selected as the one with the largest sum of squared track transverse momenta ($\sum p_T^2$). Two parameters relevant to the vertex reconstruction are illustrated in Fig. 3.8—the transverse and longitudinal impact parameters d_0 and z_0. They are defined as the distance between the interaction point and the *perigee*, the point of the track's closest approach to the beam axis, projected in the transverse and longitudinal planes.

Calorimeter Clusters Calorimeter clusters are the final detector-level objects described in this section. Two different kinds of algorithms are used for defining clusters of calorimeter cells [42]. First, a *sliding-window* algorithm can be used, which clusters calorimeter cells within fixed-size rectangles. The window's position is adjusted such that the transverse energy contained in the window is at a local maximum. The use of a fixed-size window allows for a precise energy calibration.

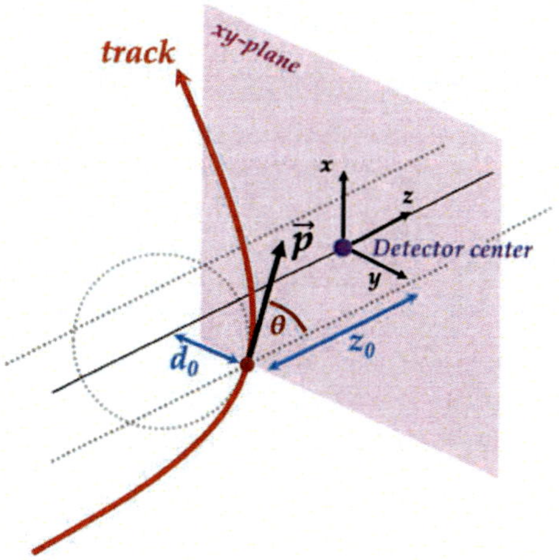

Fig. 3.8 Transverse d_0 and longitudinal z_0 impact parameters of a track with respect to the interaction point

Therefore, this algorithm is primarily used to reconstruct electrons, photons, and hadronic τ-lepton decays. Second, a topological clustering algorithm can be used, which clusters together neighbouring cells into *topo-clusters* as long as the signal is significant compared to the noise. This algorithm has the advantage of suppressing noise in clusters with a large number of cells. Therefore, it is used mainly for the reconstruction of jets and $E_{\mathrm{T}}^{\mathrm{miss}}$.

3.3.2 Electrons

The first high-level physics objects considered in this section are electrons. Electrons are identified in the central region of the detector ($|\eta| < 2.5$) by combining tracks from the ID and deposits in the electromagnetic calorimeter [42, 43]. The forward region ($2.5 < |\eta| < 4.9$) is not covered by the inner detector. Here only calorimeter information is used, but additional properties like the shape of the electromagnetic shower can be used in the electron identification. Only information about the reconstruction of central electrons is provided in the following. An additional distinction has to be made between *prompt* and *non-prompt* electrons. The former originates from the primary vertex, while the latter is produced in decays or further interactions in the detector (i.e. in the decay of the heavy flavour jets or photons producing an e^+e^- pair). Here, the identification of prompt electrons is described. The signature of electrons can be faked, for example, by mis-reconstructed photons and pions, thus it is important to have identification and isolation criteria to select electrons with high efficiency and purity.

Reconstruction Electrons are reconstructed from calorimeter clusters and track information. Tracks are matched to clusters by selecting tracks compatible with the expected energy loss of electrons in the detector. The matching is performed based on the extrapolated position of a track in the calorimeter and the barycenter of a cluster. Tracks that match a cluster are refitted using the *Gaussian Sum Filter* [44], to take into account electron energy loss via bremsstrahlung. Then, the matching of track and cluster is performed again using more stringent conditions. An electron candidate is identified when at least one track is matched to a calorimeter cluster, and in the case of multiple tracks, the track closest to the cluster is chosen. Electron candidates with no precision hits are removed and considered to be photons. Finally, for a prompt electron candidate, the tracks must be compatible with the primary vertex and satisfy additional impact parameter requirements.

Identification After the reconstruction, an additional selection is performed to identify prompt electrons and reduce background from photon conversion and non-prompt electrons. The electron identification algorithm uses a likelihood-based approach [45]. Several parameters are combined into an MVA: the number of hits in each layer, the transverse and longitudinal impact parameters, the momentum lost in the ID and information about the calorimeter cluster. Based on this, signal and background probabilities are defined, and a likelihood discriminant is built. Several operating points are defined using fixed values of the likelihood discriminant: *VeryLoose*, *Loose*, *Medium* and *Tight*, where the operating points get progressively more selective. Each operating point has a different reconstruction efficiency and background rejection, which is also dependent on p_T and $|\eta|$.

Isolation To further reduce the background for prompt electrons, electron candidates are required to be isolated from additional tracks or energy deposits, as there is typically little activity around prompt electrons. Isolation criteria can be defined based on the ID tracks and calorimeter deposits. For the track-based isolation criterion, the sum of transverse momenta of tracks with $p_T > 1$ GeV in a given ΔR cone is calculated, excluding the electron candidate. The calorimeter-based isolation criterion is defined based on the sum of energy deposits in a ΔR cone in the calorimeter, excluding the electron candidate cluster. Cuts can be applied to the two isolation criteria to provide different operating points, depending on the requirements of a physics analysis.

Calibration The total efficiency of correctly identifying an electron in the ATLAS detector is a combination of the reconstruction, identification and isolation efficiencies. The overall electron efficiency is a critical input into physics analyses. The electron efficiency on data and simulation is determined in dedicated measurements of $Z \rightarrow e^+e^-$ and $J/\psi \rightarrow e^+e^-$ events. The data and simulation measurements are compared, and scale factors are derived, which can be applied to MC samples to account for mis-modelling. The scale factors are usually close to unity [46].

3.3.3 Photons

The reconstruction and identification of photons in the ATLAS detector is intertwined with electron identification. Photons leave showers in the electromagnetic calorimeter, with no associated track. Prompt photons are reconstructed in ATLAS for $|\eta| < 2.5$ following a similar strategy as for electrons [47].

Reconstruction The reconstruction of electrons and photons proceeds in parallel, starting from electromagnetic calorimeter clusters. The candidates are then identified as either electrons, as described above, *converted* photons, where the photon converts into an e^+e^- pair, and *unconverted* photons. The distinction is made based on the presence of hits in the ID and conversion vertices. Photons with E_T >25 GeV are correctly reconstructed as photons 96% of the time and misidentified as electrons otherwise.

Identification It is essential to distinguish prompt photons against background photons (e.g. from hadronic jets). Identification criteria are defined based on the characteristics of the showers, e.g. their shape. Prompt photon showers are generally narrower than photons in hadronic jets and usually have a smaller leakage in the hadronic calorimeter. The photon identification is further complicated by pion decays to photons and low E_T activity in the detector. As for electrons, operating points are defined for photon identification based on different selection criteria.

3.3.4 Muons

Muons in the ATLAS detector are reconstructed using information from all subdetectors. They leave tracks in the ID, deposit a small amount of energy in the calorimeter (where they act as minimum ionising particles) and leave hits in the MS chambers [48]. Muons are reconstructed within the acceptance of the MS, i.e. $|\eta| < 2.7$.

Reconstruction Muons are reconstructed by first reconstructing fragments in each sub-detector and then combining the information from all components to make muon track candidates. The ID tracks are reconstructed as described in Sect. 3.3.1. In the MS, the tracks are built from segments of hits in each set of muon chambers. A fit to the hits in the MS is performed, and the muon track is selected, taking into account properties like the multiplicity and relative position of hits. The ID and MS information can then be combined, and calorimeter information can also be included to define different reconstructed muons:

- *Combined muons* are reconstructed by combining tracks from the ID and the MS and performing a combined fit to both sub-detectors to identify the combined track. Usually, the fit is performed by starting from the MS and extrapolating to the ID. Combined muons are the most commonly reconstructed muons and have the highest purity.

- *Segment-tagged muons* are reconstructed from an ID track that has at least one matching segment in the MS. This type of muon usually has low momentum or leaves the acceptance of the MS on its trajectory.
- *Calorimeter-tagged muons* combine an ID track with a calorimeter energy deposit. This type of reconstructed muon is used when the MS information is missing due to a gap in coverage or in a region where the MS has poor performance (e.g. in the central region where the MS performance is degraded by more material due to cables and support structures).
- *Extrapolated muons* use only the MS information and extrapolate to the rest of the detector. Here, at least two or three hits in the MS are required. This kind of reconstructed muon is primarily used outside of the detector acceptance of the ID ($2.5 < |\eta| < 2.7$).

Identification Identification requirements for muons are defined to distinguish prompt muons from background ones, i.e. those produced in decays of pions, kaons or heavy flavour hadrons. A typical characteristic of background muons is a kink in the reconstructed track where the decay occurred. Different variables can be used to place a requirement on the quality of a muon track. For example, it is possible to calculate the compatibility of the track momentum measured in the ID and the MS. Based on such requirements, different operating points can be required: *Loose*, *Medium*, *Tight* and *High-*p_T. These working points get increasingly more selective, which means that the purity increases, but the efficiency decreases (from 98.1% for the *loose* working point to 89.9% for the *tight* working point).

Isolation Further criteria are applied to distinguish prompt muons from background muons. Isolation criteria can be constructed using track and calorimeter information. For the track-based isolation variable, the sum of track transverse momenta in a ΔR cone around the muon candidate is used, with the cone defined using $\Delta R = \min(10\ \text{GeV}/p_T^{\mu}, 0.3)$, where p_T^{μ} is the muon transverse momentum. For the calorimeter-based isolation variable E_T^{cone20}, the sum of the energy clusters in a ΔR cone with $\Delta R = 0.2$ around the muon candidate is calculated. Different selection criteria can be constructed based on the two isolation variables to provide operating points with different efficiencies.

Calibration The reconstruction and isolation efficiencies of muons are measured in data and simulation with $J/\psi \to \mu^+\mu^-$ and $Z \to \mu^+\mu^-$ events. Scale factors are computed by comparing the efficiencies derived on data and simulated samples, which can then be applied to the simulated samples to correct for mis-modelling.

3.3.5 Jets

The most frequently produced particles in pp collisions in the LHC are quarks and gluons. Due to colour confinement, they can not be seen individually, but instead, they immediately undergo fragmentation and hadronisation, where energetic hadrons

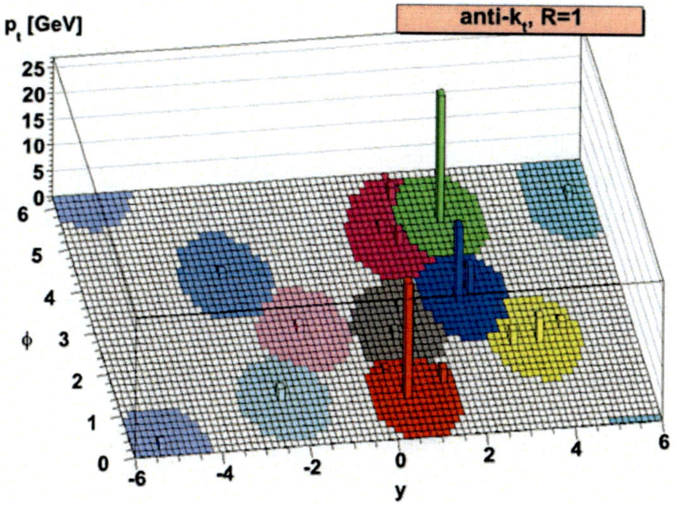

Fig. 3.9 Simulation of jet reconstruction with the anti-k_T algorithm for a HERWIG generated parton-level event. From [50]

are generated [49]. A quark or gluon generates streams of these particles, and these so-called *jets* then leave large energy deposits in the hadronic calorimeter. Jets are one of the most complicated objects in ATLAS. Ideally, the jet reconstruction should provide information about the jet momentum, direction and the type of parton which originated the jet.

Reconstruction Topo-clusters, which were discussed in Sect. 3.3.1, can be combined into jets using different jet reconstruction algorithms. In the ATLAS experiment, the anti-k_t algorithm is used [50]. The first property of this algorithm is that it satisfies *infrared and collinear (IRC) safety*. Therefore, if the event is modified with either collinear splitting or soft emission, the set of jets found by the algorithm remains the same [49].

The anti-k_T algorithm begins the reconstruction with hard (high momentum) objects and then adds soft objects to the cluster. It provides reconstructed jets circular in the $y - \phi$ direction, as can be seen in Fig. 3.9. The algorithm uses two distance parameters when considering whether to combine two cluster objects i and j:

$$d_{ij} = \min\left(\frac{1}{k_{ti}^2}, \frac{1}{k_{tj}^2}\right)\frac{\Delta R_{ij}^2}{R^2}, \qquad d_{iB} = \frac{1}{k_{ti}^2}$$

where $\Delta R_{ij}^2 = (y_i - y_j)^2 + (\phi_i - \phi_j)^2$ is the angular distance of the two objects, k_{ti} is the transverse momentum, y_i is the rapidity and ϕ_i is the azimuthal angle of object i. The parameter R, known as the radius, defines the maximum size of the jet cone. The distance between a cluster and the beam axis in momentum space is denoted as d_{iB}. The algorithm compares d_{ij} and d_{iB}. If $d_{ij} < d_{iB}$ the two objects i and j are combined and the algorithm continues to another object. If $d_{ij} > d_{iB}$ the object i is identified as a jet and removed from the list of objects considered by the algorithm. The distances between objects are then recalculated, and the process is repeated until there are no objects left. The radius parameter R can be varied depending on the purpose of the jet reconstruction. Values of $R = 0.4$ and $R = 1.0$ are typically used. $R = 0.4$ is used for jets initiated by quarks and gluons (so-called *small jets*), and $R = 1.0$ is used for hadronic decays of massive particles, like the W, Z, Higgs or top quark, or in the case of jets with high momenta (so-called *fat jets*). In the following, the focus is on *small jets*, used in the $VH(H \to c\bar{c})$ analysis described in Chap. 4.

Calibration Once reconstructed, the energy of the jets has to be adjusted to account for several effects: detector effects like calorimeter non-compensation, the difference between the electromagnetic and hadronic calorimeter, dead material and leakage of the jet outside of the calorimeter, as well as reconstruction effects, such as energy deposited outside of the jet cone, and effects related to pile-up. Additionally, the *Jet Energy Scale (JES)* has to be adjusted to match the scale of truth-jets, i.e. the jets in simulated samples. The JES calibration is done in several steps, using data-driven techniques and simulated samples [51]:

1. **Origin correction** During the initial jet reconstruction, the jet axis points to the detector centre. In this step, the direction of the jet is adjusted, such that the jet axis points towards the primary vertex in the event. The four-momentum is recalculated, while the jet energy is kept constant.
2. **Pile-up corrections** Due to the *in-time* and *out-of-time* pile-up present in the collisions, additional energy deposits in the calorimeter can increase the energy attributed to a jet. The goal is to remove these pile-up contributions to the jet energy in two steps [52]. First, an overall correction is subtracted from the jet p_T, based on the average pile-up contribution. Then, a second correction is applied based on the number of interactions per bunch crossing (μ) and the number of reconstructed primary vertices in the event.
3. **Absolute JES correction** The energy of the reconstructed jets is adjusted from the reconstruction scale to the particle-level energy scale used in simulated events. A correction is applied based on comparing the reconstructed jets (with the previous corrections applied) in data and truth jets, derived with a geometrical matching of $\Delta R = 0.3$. The corrections are derived and applied as a function of E and η.
4. **Eta inter-calibration** Jets in the forward region ($0.8 < |\eta| < 2.5$) are usually less well-measured, so an additional correction is applied to these jets based on the well-measured jets in the central region ($|\eta| < 1.4$).
5. **Global Sequential Calibration (GSC)** The energy leakage in the calorimeters can cause fluctuations in the jet response. An additional correction is applied to account for this. Five observables are defined, which depend on the shape and energy of the jet. For each observable a four-momentum correction is derived,

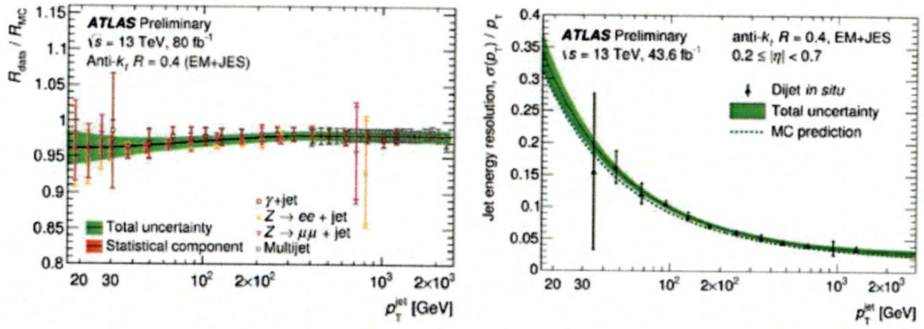

Fig. 3.10 Data-to-simulation ratio of the average jet p_T response as a function of jet p_T for JES (left). From [54]. The relative jet energy resolution σ_{p_T}/p_T as a function of p_T for anti-k_T jets with a radius parameter of R = 0.4 and inputs of EM-scale topoclusters. From [55]

depending also on the jet p_T^{truth} and $|\eta|$. The GSC improves the distinction between gluon and quark initiated jets.

6. **In-situ calibration** Finally, a correction is applied to account for possible differences between data and simulation because of an incorrect detector description in simulation. The JES is calibrated in a dedicated measurement in data, using a well-calibrated object as a reference, by calculating the p_T balance with respect to that object and performing a Gaussian fit to extract the correction. At low momenta ($20 < p_T < 500$ GeV), $Z+$jet events are used, where the Z decays leptonically. At intermediate momenta ($36 < p_T < 900$ GeV), $\gamma+$jet events are used. QCD multi-jet events are used at very high momenta (up to 2 TeV), where a multi-jet hadronic recoil is used as the reference object. Here, the momentum balance of a high p_T jet is computed against several jets with lower p_T, which are generally calibrated better.

After performing the calibration, the JES and its related uncertainties are evaluated as a function of jet p_T. Following the JES calibration, the *jet energy resolution (JER)* can be defined as σ_{p_T}/p_T, and is derived and calibrated *in-situ* in di-jet events [53]. The results are shown in Fig. 3.10.

Both the JES and JER are evaluated with a large number of systematic uncertainties. These are mainly related to the data-driven calibration stages, e.g. the available statistics, assumptions made on the simulated samples, mis-modelling of pile-up or the jet flavour composition in MC and uncertainties on the energy scales of other objects. The systematic uncertainties are assessed by varying parameters of the object selection and using alternative simulated samples. The total uncertainty on the JES calibration is 4.5% at 20 GeV and reduces to a few percent above 200 GeV in di-jet events. The uncertainties in the JER calibration vary between 3% at 20 GeV and less than 1% above 40 GeV in a di-jet sample.

Suppression of pile-up jets Removing or suppressing pile-up jets is crucial for physics analyses. Even with the pile-up corrections applied to the jet energy scale, it is still possible to see the effects of pile-up jets due to fluctuations in pile-up activity.

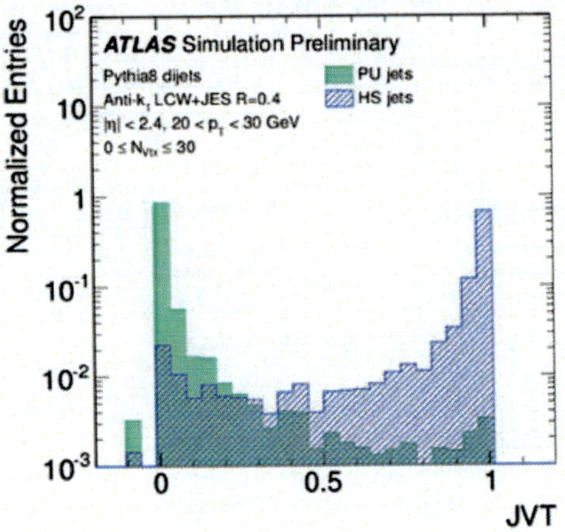

Fig. 3.11 Distribution of JVT for pile-up (PU) and hard-scatter (HS) jets with $20 < p_T < 30$ GeV. From [56]

To specifically remove these jets, the *Jet Vertex Tagger (JVT)* is constructed [56]. A two-dimensional likelihood is built from variables based on tracking information (e.g. the weighted fraction of the transverse momentum of tracks within the jet originating from the hard scatter). The resulting JVT discriminant is shown in Fig. 3.11. Based on this discriminant, selection criteria can be constructed to define different working points, with particular efficiencies and pile-up fake rates.

3.3.6 Flavour Tagging of Jets

Identifying the flavour of a quark from which a jet originated is very important in physics analyses to fully reconstruct the hard scatter process. In ATLAS *flavour tagging* (the flavour identification of jets) is specially developed for the identification of b-quark jets and has been crucial to many analyses, for example, the observation of $VH(H \rightarrow b\bar{b})$ decays. More recently, flavour tagging has also been used for the identification of c-quark jets, for example, in searches for $VH(H \rightarrow c\bar{c})$ decays, as described in Chap. 4. This section outlines the basic principles of flavour tagging methodologies used in the ATLAS experiment and focuses on the methods developed for b-jet identification. The application of these methods to c-jets in the context of the $VH(H \rightarrow c\bar{c})$ analysis is discussed in Sect. 4.5.

In flavour tagging, a distinction is made between b-jets, c-jets and light-jets, where the latter include jets originating from u, d, s-quarks and gluons [57]. Jets originating from hadronic decays of τ-leptons are also considered. To understand the flavour tagging algorithms used, it is important to understand the characteristics of the different kinds of jets, which are illustrated in Fig. 3.12.

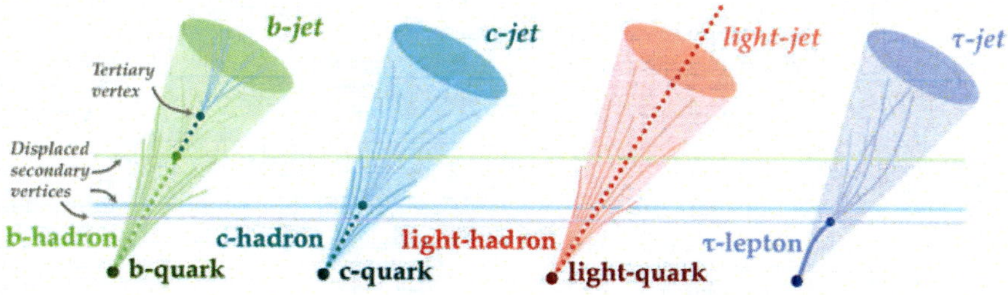

Fig. 3.12 Illustration of jets initiated by different kinds of quarks and leptons

- ***b*-jets** originate from the hadronisation of a b-quark into a b-hadron. The relatively long lifetime (1.5 ps) of b-hadrons means that the hadron travels an average distance of around 5 mm inside the detector (depending on its momentum) before it decays. This means that a second displaced vertex can be found inside the jet, as illustrated in Fig. 3.12. The b-hadron carries approximately 75% of the jet energy and has a mass of 5 GeV, which results in harder objects inside the jet and a larger decay multiplicity compared to other flavours of jets. Most b-hadrons (∼90%) decay into c-hadrons, due to the CKM matrix transition values $|V_{cb}|^2 > |V_{ub}|^2$ [58]. The c-hadrons then decay further after travelling a few mm in the detector, which means a tertiary vertex could also be present inside of the b-jet.
- ***c*-jets** originate from the hadronisation of a c-quark. Usually, a D-meson is produced, which most commonly decays to a kaon via the $c \rightarrow s$ transition. D-mesons have a shorter lifetime than b-hadrons (between 0.2 and 1 ps). Therefore, the D-meson's decay length is shorter and it only travels a few mm inside of the detector. However, the secondary vertex is still measurable. The D-meson usually carries around 55% of the jet energy, which leads to a softer jet, and has a mass of around 2 GeV, which results in a lower multiplicity of decay products. The difference in jet energy carried by b-, c- and light-hadrons is caused by the dead-cone effect, which suppresses gluon radiation for heavier quarks [59].
- **light-jets** originate from the hadronisation of u, d, s-quarks or gluons. The quarks hadronise into light-hadrons almost instantly, and several hadrons share the jet energy. Long-lived hadrons can be produced, but these decay very far from the primary vertex. Hence, the secondary vertex can not be reconstructed in the ID. Gluon jets will also not have a distinct secondary vertex. They can be distinguished from quark jets due to a larger number of jet constituents and broader radiation patterns [60], but are considered within the light-jet category in the context of this thesis.
- ***τ*-jets** originate from the hadronic decays of the τ-leptons. The τ-lepton has a mass of 1.8 GeV and a lifetime of 0.3 ps [61], which means that τ-jets usually also have a displaced vertex and share many characteristics with c-jets, though τ-jets are generally narrower. The specific reconstruction of hadronic τ-leptons is presented in Sect. 3.3.7, and the implications of τ-jets in c-tagging are discussed in Chap. 4.

Within the ATLAS experiment, different kinds of flavour tagging algorithms are used. So-called *low-level taggers* try to extract features relevant to discriminating different types of jets using the tracks in the ID and the associated jets, e.g. the location of vertices and the impact parameters of tracks. For these algorithms, the matching of ID tracks to their associated jets is performed based on the angular separation of the jet and track $\Delta R(\text{jet}, \text{track})$. The ΔR requirement is defined as a function of jet p_T to account for the collimation of decay products at high momenta [62, 63]. The output of these low-level taggers then serves as input to the *high-level taggers*, which use machine learning algorithms to extract a probability of a jet being of a particular flavour.

Impact-Parameter Based Algorithms
The first low-level flavour tagging algorithms utilise information about the impact parameters of the tracks in the ID. The transverse and longitudinal impact parameters have been defined in Sect. 3.3.1, and due to their lifetime, tracks from b- and c-hadrons are expected to have larger impact parameters compared to tracks originating from the primary vertex. ATLAS has two impact parameter based algorithms, `IP2D` and `IP3D`, which utilise the signed impact parameter significance [64, 65]. The sign of the impact parameter is defined by whether the point of closest approach of the track to the PV is in front (positive) or behind (negative) the PV, with respect to the direction of the jet. By definition, this means that tracks associated with b- and c-hadrons are more likely to have a positive impact parameter. The impact parameter significance is defined as the ratio of the impact parameter and its uncertainty and can be calculated in the transverse (d_0/σ_{d_0}) and longitudinal ($z_0 \sin\theta/\sigma_{z_0}$) planes. The signed impact parameters for different flavours of jets in simulated $t\bar{t}$ events are shown in Fig. 3.13. Using the probability density function of these quantities, the `IP2D` and `IP3D` algorithms construct a ratio of the b-jet and light-jet probabilities for each of the tracks, which are then combined into a log-likelihood discriminant. The `IP2D` algorithm only uses the transverse impact parameter, whereas the `IP3D` algorithm uses both the transverse and longitudinal impact parameter and takes into account their correlation.

Secondary Vertex Based Algorithms
The secondary vertex-finding algorithm `SV1` aims to reconstruct a single displaced vertex inside of the jet [66]. The secondary vertex reconstruction is performed in multiple steps. First, two-track vertices are constructed from the candidate tracks while trying to reject tracks from background processes, such as decays of long-lived particles, photon conversion or hadronic interactions with the detector material. As a second step, all tracks are combined into one set, and the fitting procedure tries to fit the secondary vertex (SV) by accepting or rejecting tracks from the vertex iteratively, based on a χ^2 test [67, 68]. Once the secondary vertex is identified, a set of variables is constructed, which can be used in the high-level taggers. The variables are related to the properties of the secondary vertex, for example, the reconstructed mass at the SV, the ratio of energy associated with the SV compared to all tracks in the jet, or the number of tracks used to reconstruct the SV. Examples of the SV variables are shown in Fig. 3.14.

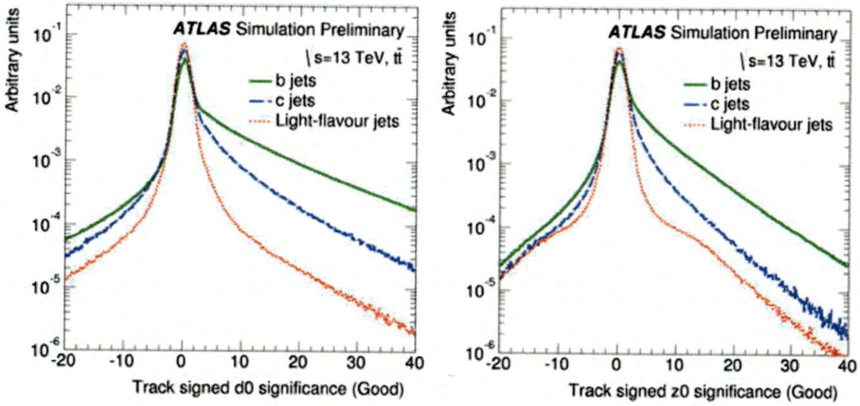

Fig. 3.13 **The transverse (left) and longitudinal (right) signed impact parameter significance of** tracks in $t\bar{t}$ events associated with b- (solid green), c- **(dashed blue) and light-flavour (dotted red)** jets. The distributions are shown for tracks in the *good* **category, defined by requirements such as** no missing hits in layer 0 and 1 and no shared pixel hits. From [65]

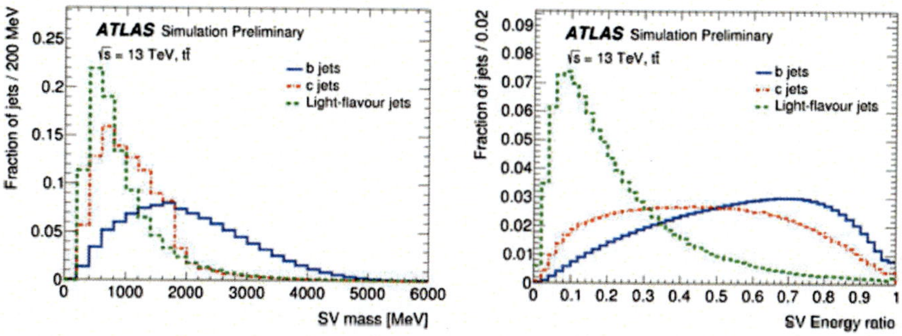

Fig. 3.14 Secondary vertex mass (left) and energy (right) distributions with b-jets (blue), c-jets (red) and light jets (green). All distributions have been normalised to one. From [66]

Decay Chain Multi-vertex Algorithm: `JetFitter`

The decay chain multi-vertex algorithm `JetFitter` exploits the topological structure of b- and c-hadron decays by reconstructing the complete hadron decay chain [69, 70]. This means the algorithm tries to reconstruct the complete decay process from b-hadron to c-hadron to light-hadron and the associated secondary and tertiary vertices. This is achieved by employing a modified Kalman filter that attempts to find a common line on which the secondary and tertiary vertices lie and their locations on the line. Based on this result, a set of variables related to the multiple vertices is constructed, which can then be used by the higher-level algorithms.

Higher-Level Flavour Tagging Algorithms

Finally, the outputs of the low-level algorithms are combined into high-level algorithms. In the context of this thesis, two of these algorithms are discussed: `MV2c10` and DL1 [71, 72], which are relevant to the $VH(H \rightarrow c\bar{c})$ analysis. Table 3.1 shows the input variables used in the high-level taggers. They include variables from IP2D, IP3D, SV1 and JetFitter. In the case of the DL1 algorithm, additional JetFitter c-tagging variables are used, which include some additional properties related to the secondary and tertiary vertices. Both algorithms are trained on a *hybrid* sample of $t\bar{t}$ and Z' events, where the $t\bar{t}$ events are selected to provide a large number of b-jets, whereas the Z' sample is generated to provide a uniform sample of all jet flavours up to high p_T.

MV2c10 The MV2c10 algorithm is a flavour tagging algorithm developed explicitly for the tagging of b-jets. Using b-jets as the signal and c- and light-jets as the background, a Boosted Decision Tree (BDT) is trained. A plot of the resulting discriminant is shown in Fig. 3.15 (left). Different operating points can be defined using cuts on the discriminating variable, which have different efficiencies of identifying b-jets. In ATLAS, four working points are derived, with efficiencies of 60%, 70%, 77% and 85%, measured on simulated $t\bar{t}$ events.

DL1 The DL1 algorithm is trained as a deep feed-forward neural network. It provides a multidimensional output, which corresponds to the probabilities of a particular jet being a b-, c- or light-jet. As a consequence, the DL1 output can be used for both b- and c-tagging, where the application to c-tagging is discussed in Chap. 4. For b-tagging, the following discriminant can be constructed:

$$DL_1 = \ln \frac{p_b}{f_c p_c + (1 - f_c) p_l} \tag{3.5}$$

where p_b, p_c and p_l are the b-, c- and light-jet probabilities provided by DL1, and f_c is a weight applied to the c- and light-jets. The f_c parameter can be optimised, but for b-tagging in ATLAS it is set to $f_c = 0.08$. The resulting DL1 b-tagging discriminant is shown in Fig. 3.15 (right). Again, cuts can be applied to the DL1 discriminant to define different working points for flavour tagging.

For the high-level taggers, the efficiency of correctly identifying a b-jet as such can be determined in simulation samples, defined as:

$$\epsilon_b(p_T) = \frac{N^{\text{b-jets}}_{\text{passing selection}}(p_T)}{N^{\text{b-jets}}_{\text{total}}(p_T)} \tag{3.6}$$

where $N^{\text{b-jets}}_{\text{passing selection}}(p_T)$ is the number of b-jets that are b-tagged and $N^{\text{b-jets}}_{\text{total}}(p_T)$ is the total number of b-jets in the sample. The efficiency is often given as a function of jet momentum. Similar efficiencies can be defined for the misidentification of c- and light-jets, and the c- and light-flavour *rejection* is defined as $1/\epsilon_c$ and $1/\epsilon_{light}$. Figure 3.16 shows the c- and light-rejection of the low-level and high-level taggers as a function of b-tagging efficiency. There is a significant benefit from combining all of the information from the low-level taggers into the high-level taggers. Of note is that the DL1 algorithm provides better c- and light-rejection for the same b-tagging efficiencies, compared to the MV2c10 algorithm.

Table 3.1 Input variables used by the MV2 and the DL1 algorithms. The JetFitter c-tagging variables are only used by the DL1 algorithm. From [71]

Input	Variable	Description
Kinematics	p_T	Jet p_T
	η	Jet $\lvert\eta\rvert$
IP2D/IP3D	$\log(P_b/P_{\text{light}})$	Likelihood ratio between the b-jet and light-flavour jet hypotheses
	$\log(P_b/P_c)$	Likelihood ratio between the b- and c-jet hypotheses
	$\log(P_c/P_{\text{light}})$	Likelihood ratio between the c-jet and light-flavour jet hypotheses
SV1	$m(\text{SV})$	Invariant mass of tracks at the secondary vertex assuming pion mass
	$f_E(\text{SV})$	Energy fraction of the tracks associated with the secondary vertex
	$N_{\text{TrkAtVtx}}(\text{SV})$	Number of tracks used in the secondary vertex
	$N_{\text{2TrkVtx}}(\text{SV})$	Number of two-track vertex candidates
	$L_{xy}(\text{SV})$	Transverse distance between the primary and secondary vertex
	$L_{xyz}(\text{SV})$	Distance between the primary and the secondary vertex
	$S_{xyz}(\text{SV})$	Distance between the primary and the secondary vertex divided by its uncertainty
	$\Delta R(\vec{p}_{\text{jet}}, \vec{p}_{\text{vtx}})(\text{SV})$	ΔR between the jet axis and the direction of the secondary vertex relative to the primary vertex.
JetFitter	$m(\text{JF})$	Invariant mass of tracks from displaced vertices
	$f_E(\text{JF})$	Energy fraction of the tracks associated with the displaced vertices
	$\Delta R(\vec{p}_{\text{jet}}, \vec{p}_{\text{vtx}})(\text{JF})$	ΔR between the jet axis and the vectorial sum of momenta of all tracks attached to displaced vertices
	$S_{xyz}(\text{JF})$	Significance of the average distance between PV and displaced vertices
	$N_{\text{TrkAtVtx}}(\text{JF})$	Number of tracks from multi-prong displaced vertices
	$N_{\text{2TrkVtx}}(\text{JF})$	Number of two-track vertex candidates (prior to decay chain fit)
	$N_{\text{1-trk vertices}}(\text{JF})$	Number of single-prong displaced vertices
	$N_{\geq\text{2-trk vertices}}(\text{JF})$	Number of multi-prong displaced vertices
JetFitter c-tagging	$L_{xyz}(\text{2nd/3rd vtx})(\text{JF})$	Distance of 2nd or 3rd vertex from PV
	$L_{xy}(\text{2nd/3rd vtx})(\text{JF})$	Transverse displacement of the 2nd or 3rd vertex
	$m_{\text{Trk}}(\text{2nd/3rd vtx})(\text{JF})$	Invariant mass of tracks associated with 2nd or 3rd vertex
	$E_{\text{Trk}}(\text{2nd/3rd vtx})(\text{JF})$	Energy fraction of the tracks associated with 2nd or 3rd vertex
	$f_E(\text{2nd/3rd vtx})(\text{JF})$	Fraction of charged jet energy in 2nd or 3rd vertex
	$N_{\text{TrkAtVtx}}(\text{2nd/3rd vtx})(\text{JF})$	Number of tracks associated with 2nd or 3rd vertex
	$Y_{\text{trk}}^{\text{min}}, Y_{\text{trk}}^{\text{max}}, Y_{\text{trk}}^{\text{avg}}(\text{2nd/3rd vtx})(\text{JF})$	Min., max. and avg. track rapidity of tracks at 2nd or 3rd vertex

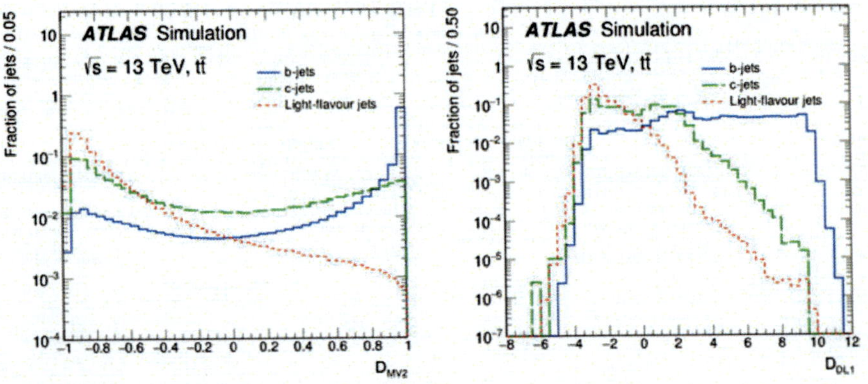

Fig. 3.15 Distribution of the output discriminant of the MV2 (left) and DL1 (right) b-tagging algorithms for b-jets, c-jets and light-flavour jets in $t\bar{t}$ simulated events. From [71]

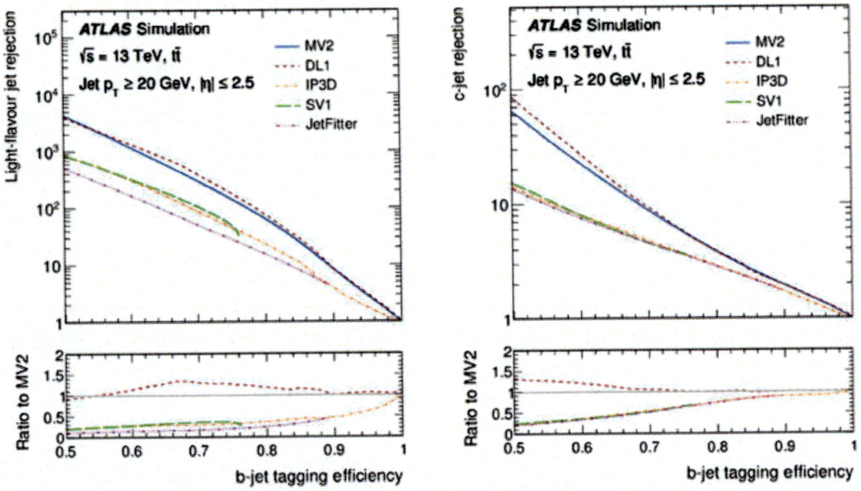

Fig. 3.16 The light-flavour jet (left) and c-jet rejections (right) versus the b-jet tagging efficiency for the IP3D, SV1, JetFitter, MV2 and DL1 b-tagging algorithms evaluated on $t\bar{t}$ events. From [71]

After its definition, the flavour tagging working point has to be calibrated. Data-to-simulation scale factors are derived as a function of jet momentum by comparing the performance of the flavour tagger in simulated samples and data. These scale factors are then applied to simulated samples when the flavour tagging algorithm is used in a physics analysis. The calibration is performed separately for each jet flavour and is discussed further for the c-tagger used in the $VH(H \rightarrow c\bar{c})$ analysis in Sect. 4.5.

3.3.7 Tau-Leptons

Tau-leptons ($\tau^{\pm}$) are the heaviest leptons, with a mass of 1.78 GeV [61]. They have a very short lifetime, which results in a small average distance travelled within the ATLAS detector ($\langle c\tau\beta\gamma\rangle \sim 5$ mm) before the τ-lepton decays inside of the LHC beampipe. The τ-lepton decays leptonically 35% of the time - into two neutrinos and either an electron or a muon. The electrons and muons produced in these decays are very difficult to distinguish from prompt leptons, so the reconstruction of leptonically decaying τ-leptons is very challenging. The τ reconstruction in ATLAS focuses on the hadronic decay modes. These decays typically contain either one charged pion (75% of the time, *1-prong* signature) or three charged pions (22% of the time, *3-prong* signature), as well as occasionally one or two neutral pions which decay as $\pi^0 \rightarrow \gamma\gamma$. The hadronic decays can be identified as narrow jets in the calorimeter, reconstructed as *small jets* ($R = 0.4$), with some associated tracks in the ID [73]. To further distinguish between τ-jets and QCD jets, the characteristics of τ-jets are exploited, e.g. the low track multiplicity, collimated energy deposits and the identification of a τ decay vertex. A set of variables relevant to τ-jets is used to train a BDT on a sample of $Z/\gamma \rightarrow \tau\tau$ events, separately for *1-prong* and *3-prong* decays [74]. Based on the BDT output, a set of τ-lepton identification working points can be defined with different efficiencies.

3.3.8 Missing Transverse Momentum and Energy

So far, all the discussed physics objects originated from particles that interact with the ATLAS detector and have distinct signatures. However, some particles do not leave significant energy deposits in the ATLAS detector and cannot be measured directly, such as neutrinos and potential particles beyond the Standard Model (e.g. dark matter candidates). The effect of these particles is quantified by considering the momentum imbalance in the transverse plane of the detector. In pp collisions, the initial momentum of the partons along the beam-axis is unknown, but in the transverse plane, it has to be zero. Any momentum imbalance in the transverse plane is defined by the quantity $\vec{E}_{\mathrm{T}}^{\mathrm{miss}}$, which is the vectorial sum:

$$\begin{aligned} \vec{E}_T^{miss} &= -\Big(\sum_{hard} \vec{p}_T + \sum_{soft} \vec{p}_T\Big) \\ &= -\Big(\sum_{e} \vec{p}_T + \sum_{\mu} \vec{p}_T + \sum_{\tau} \vec{p}_T + \sum_{\gamma} \vec{p}_T + \sum_{jets} \vec{p}_T + \sum_{soft} \vec{p}_T\Big) \end{aligned} \tag{3.7}$$

and is comprised of a *soft* and a *hard* term [75]. The *hard* term includes the momenta of all of the fully reconstructed objects. In contrast, the *soft* term takes into account the charged particle tracks in the ID, which are associated with the primary vertex but not with any other object. The *hard* term is calculated using the existing physics objects, e.g. electrons or muons, reconstructed using the methodologies described above. Additionally, the physics objects have to pass the event selection of the particular analysis in which they are used, and further optimisation can be performed within a physics analysis to improve the $\vec{E}_{\mathrm{T}}^{\mathrm{miss}}$ reconstruction. The *soft* term is reconstructed from all of the remaining ID tracks associated with the primary vertex. This provides an important contribution to the $\vec{E}_{\mathrm{T}}^{\mathrm{miss}}$ calculation, in particular in analyses with few hard objects. The performance of the $\vec{E}_{\mathrm{T}}^{\mathrm{miss}}$ reconstruction is measured in data and simulation using events with zero expected $E_{\mathrm{T}}^{\mathrm{miss}}$ (i.e. $Z \rightarrow \ell\ell$), and in events where $\vec{E}_{\mathrm{T}}^{\mathrm{miss}}$ is expected from at least one neutrino (i.e. W or top quark decays). The scale and resolution of the reconstructed $\vec{E}_{\mathrm{T}}^{\mathrm{miss}}$ is determined, along with its related uncertainties, which must be taken into account by physics analyses.

References

1. Brüning OS et al (2004) LHC design report. CERN yellow reports: monographs. Geneva: CERN. https://doi.org/10.5170/CERN-2004-003-V-1. https://cds.cern.ch/record/782076
2. The ALICE Collaboration (2008) The ALICE experiment at the CERN LHC. J Instrum 3(08):S08002–S08002. https://doi.org/10.1088/1748-0221/3/08/s08002
3. ATLAS Collaboration (2008) The ATLAS experiment at the CERN large hadron collider. J Instrum 3(08):S08003–S08003. https://doi.org/10.1088/1748-0221/3/08/s08003
4. The CMS Collaboration (2008) The CMS experiment at the CERN LHC. J Instrum 3(08):S08004–S08004. https://doi.org/10.1088/1748-0221/3/08/s08004
5. The LHCb Collaboration (2008) The LHCb detector at the LHC. J Instrum 3(08):S08005–S08005. https://doi.org/10.1088/1748-0221/3/08/s08005
6. Mobs E (2019) The CERN accelerator complex - 2019. Complexe des accélérateurs du CERN - 2019. In: General photo. https://cds.cern.ch/record/2684277
7. Bruning OS et al (2004) LHC design report vol 1: The LHC main ring
8. Luminosity Public Results Run-2 (2019) Geneva, July 2019. https://twiki.cern.ch/twiki/bin/view/AtlasPublic/LuminosityPublicResultsRun2
9. ATLAS Collaboration (2019) Luminosity determination in pp collisions at $\sqrt{s} = 13$ TeV using the ATLAS detector at the LHC. ATLAS-CONF-2019-021. https://cds.cern.ch/record/2677054
10. ATLAS Collaboration (1997) ATLAS magnet system: technical design report, 1. Technical design report. Geneva: CERN. https://cds.cern.ch/record/338080
11. ATLAS Collaboration (2015) Track reconstruction performance of the ATLAS inner detector at $\sqrt{s} = 13$ TeV. ATL-PHYS-PUB-2015-018. https://cds.cern.ch/record/2037683
12. ATLAS Collaboration (1997) ATLAS inner detector: technical design report, 1. Technical design report. ATLAS. Geneva: CERN. https://cds.cern.ch/record/331063
13. Stanecka E (2013) The ATLAS inner detector operation, data quality and tracking performance. arXiv:1303.3630 [physics.ins-det]
14. Potamianos K (2016) The upgraded pixel detector and the commissioning of the inner detector tracking of the ATLAS experiment for Run-2 at the LargeHadron collider. Technical report Geneva: CERN. arXiv:1608.07850
15. Capeans M et al (2010) ATLAS insertable B-layer technical design report. Technical report. https://cds.cern.ch/record/1291633
16. **ATLAS Collaboration (2016) IBL efficiency and single point resolution in collision events.** Technical report Geneva: CERN. https://cds.cern.ch/record/2203893
17. Pernegger H (2015) The pixel detector of the ATLAS experiment for LHC Run-2. Technical report Geneva: CERN. https://doi.org/10.1088/1748-0221/10/06/C06012

18. ATLAS collaboration (2007) The silicon microstrip sensors of the ATLAS semiconductor tracker. Technical report Geneva: CERN. https://doi.org/10.1016/j.nima.2007.04.157
19. The ATLAS collaboration (2014) Operation and performance of the ATLAS semiconductor tracker. J Instrum 9(08):P08009–P08009. ISSN: 1748-0221. https://doi.org/10.1088/1748-0221/9/08/p08009
20. Vogel A (2013) ATLAS transition radiation tracker (TRT): straw tube gaseous detectors at high rates. Technical report Geneva: CERN. https://cds.cern.ch/record/1537991
21. Pequenao J (2008) Computer generated image of the ATLAS calorimeter. https://cds.cern.ch/record/1095927
22. Airapetian A et al (1999) ATLAS: detector and physics performance technical design report, vol 1
23. ATLAS Collaboration (1996) ATLAS liquid-argon calorimeter: technical design report. Technical design report. ATLAS. Geneva: CERN. https://cds.cern.ch/record/331061
24. Aubert B et al (2006) Construction, assembly and tests of the ATLAS electromagnetic barrel calorimeter. Nucl Instrum Meth A 558:388–418. https://doi.org/10.1016/j.nima.2005.11.212
25. Gingrich DM et al (2007) Construction, assembly and testing of the ATLAS hadronic end-cap calorimeter. JINST 2:P05005. https://doi.org/10.1088/1748-0221/2/05/P05005
26. Orr RS (2011) The ATLAS forward calorimeter. https://cds.cern.ch/record/1386649
27. ATLAS Collaboration (1996) ATLAS tile calorimeter: technical design report. Technical design report. ATLAS. Geneva: CERN. https://cds.cern.ch/record/331062
28. Ilic N (2014) Performance of the ATLAS liquid argon calorimeter after three years of LHC operation and plans for a future upgrade. J Instrum 9(03):C03049–C03049. https://doi.org/10.1088/1748-0221/9/03/c03049
29. Pequenao J (2008) Computer generated image of the ATLAS Muons subsystem. https://cds.cern.ch/record/1095929
30. ATLAS Collaboration (1997) ATLAS muon spectrometer: technical design report. Technical design report. ATLAS. Geneva: CERN. https://cds.cern.ch/record/331068
31. Diehl E (2011) Calibration and performance of the ATLAS muon spectrometer. Technical report Geneva: CERN. arXiv:1109.6933
32. Bauer F et al (2001) Construction and test of MDT chambers for the ATLAS muon spectrometer. Nucl Inst Meth A 461(1):17–20. 8th Pisa meeting on advanced detectors. ISSN: 0168-9002. https://doi.org/10.1016/S0168-9002(00)01156-6
33. Argyropoulos T et al (2009) Cathode strip chambers in ATLAS: installation, commissioning and in-situ performance. IEEE Trans Nucl Sci 56(3):1568–1574. https://doi.org/10.1109/TNS.2009.2020861
34. Aielli G et al (2006) The RPC first level muon trigger in the barrel of the ATLAS experiment. Nucl Phys B - Proc Suppl 158:11–15. Proceedings of the 8th international workshop on resistive plate chambers and related detectors. ISSN: 0920-5632. https://doi.org/10.1016/j.nuclphysbps.2006.07.031
35. Majewski S et al (1983) A thin multiwire chamber operating in the high multiplication mode. Nucl Inst Meth 217(1):265–271. ISSN: 0167-5087. https://doi.org/10.1016/0167-5087(83)90146-1
36. zur Nedden M (2017) The LHC run 2 ATLAS trigger system: design, performance and plans. J Instrum 12(03):C03024–C03024. https://doi.org/10.1088/1748-0221/12/03/c03024
37. Pequenao J, Schaffner P (2013) How ATLAS detects particles: diagram of particle paths in the detector. https://cds.cern.ch/record/1505342
38. Aaboud M et al (2017) Performance of the ATLAS track reconstruction algorithms in dense environments in LHC Run 2. Eur Phys J C 77(10):673
39. Kalman RE (1960) A new approach to linear filtering and prediction problems. J Basic Eng 82(1):35–45. ISSN: 0021-9223. https://doi.org/10.1115/1.3662552
40. Aaboud M et al (2017) Reconstruction of primary vertices at the ATLAS experiment in Run 1 proton–proton collisions at the LHC. Eur Phys J C 77(5):332
41. ATLAS Collaboration (2010) Performance of primary vertex reconstruction in proton-proton collisions at $\sqrt{s} = 7$ TeV in the ATLAS experiment. Technical report Geneva: CERN. https://cds.cern.ch/record/1281344

42. Lampl W et al (2008) Calorimeter clustering algorithms: description and performance. Technical report Geneva: CERN. https://cds.cern.ch/record/1099735
43. Aaboud M et al (2019) Electron and photon energy calibration with the ATLAS detector using 2015–2016 LHC proton-proton collision data. J Instrum 14(03):P03017–P03017. ISSN: 1748-0221. https://doi.org/10.1088/1748-0221/14/03/p03017
44. ATLAS Collaboration (2012) Improved electron reconstruction in ATLAS using the Gaussian sum filter-based model for bremsstrahlung. Technical report Geneva: CERN. https://cds.cern.ch/record/1449796
45. Aaboud M et al (2019) Electron reconstruction and identification in the ATLAS experiment using the 2015 and 2016 LHC proton–proton collision data at $\sqrt{s} = 13$ TeV. Eur Phys J C 79(8). ISSN: 1434-6052. https://doi.org/10.1140/epjc/s10052-019-7140-6
46. ATLAS Collaboration (2019) Electron reconstruction and identification in the ATLAS experiment using the 2015 and 2016 LHC proton–proton collision data at $\sqrt{s} = 13$ TeV. Eur Phys J C 79(8):639
47. Aaboud M et al (2019) Measurement of the photon identification efficiencies with the ATLAS detector using LHC Run 2 data collected in 2015 and 2016. Eur Phys J C 79(3). ISSN: 1434-6052. https://doi.org/10.1140/epjc/s10052-019-6650-6
48. ATLAS Collaboration (2016) Muon reconstruction performance of the ATLAS detector in proton-proton collision data at $\sqrt{s} = 13$ TeV. Eur Phys J C 76(5):292. https://doi.org/10.1140/epjc/s10052-016-4120-y
49. Salam GP (2010) Towards jetography. Eur Phys J C 67(3–4):637–686. ISSN: 1434-6052. https://doi.org/10.1140/epjc/s10052-010-1314-6
50. Cacciari M, Salam GP, Soyez G (2008) The anti-kt jet clustering algorithm. J High Energy Phys 2008(04):063–063. https://doi.org/10.1088/1126-6708/2008/04/063
51. ATLAS collaboration (2017) Jet energy scale measurements and their systematic uncertainties in proton-proton collisions at $\sqrt{s} = 13$ TeV with the ATLAS detector. Phys Rev D 96:072002. 36 p. https://doi.org/10.1103/PhysRevD.96.072002. arXiv:1703.09665
52. Cacciari M, Salam GP (2008) Pileup subtraction using jet areas. Phys Lett B 659(1–2):119–126. ISSN: 0370-2693. https://doi.org/10.1016/j.physletb.2007.09.077
53. Aad G et al (2013) Jet energy resolution in proton-proton collisions at $\sqrt{s} = 7$ TeV recorded in 2010 with the ATLAS detector. Eur Phys J C 73(3). ISSN: 1434-6052. https://doi.org/10.1140/epjc/s10052-013-2306-0
54. ATLAS Collaboration (2018) Jet energy scale and uncertainties in 2015-2017 data and simulation. https://atlas.web.cern.ch/Atlas/GROUPS/PHYSICS/PLOTS/JETM-2018-006/
55. ATLAS Collaboration (2018) Jet energy resolution in 2017 data and simulation. https://atlas.web.cern.ch/Atlas/GROUPS/PHYSICS/PLOTS/JETM-2018-005/
56. ATLAS Collaboration (2014) Tagging and suppression of pileup jets with the ATLAS detector. Technical report Geneva: CERN. https://cds.cern.ch/record/1700870
57. ATLAS Collaboration (2017) Optimisation and performance studies of the ATLAS b -tagging algorithms for the 2017-18 LHC run. Technical report Geneva: CERN. https://cds.cern.ch/record/2273281
58. Kobayashi M, Maskawa T (1973) CP violation in the renormalizable theory of weak interaction. Prog Theor Phys 49:652–657. https://doi.org/10.1143/PTP.49.652
59. ALICE Collaboration (2022) Direct observation of the dead-cone effect in quantum chromodynamics. Nature 605(7910):440–446. https://doi.org/10.1038/s41586-022-04572-w. https://doi.org/10.1038/s41586-022-04572-w
60. ATLAS Collaboration (2017) Quark versus gluon jet tagging using jet images with the ATLAS detector. Technical report Geneva: CERN. https://cds.cern.ch/record/2275641
61. Tanabashi M et al (2018) Review of particle physics. Phys Rev D 98:030001. https://doi.org/10.1103/PhysRevD.98.030001
62. ATLAS Collaboration (2010) Performance of the ATLAS secondary vertex b-tagging algorithm in 900 GeV collision data. Technical report Geneva: CERN. https://cds.cern.ch/record/1273194
63. ATLAS Collaboration (2011) Commissioning of the ATLAS high-performance b-tagging algorithms in the 7 TeV collision data. Technical report Geneva: CERN. https://cds.cern.ch/record/1369219

64. **ATLAS Collaboration (2016) Performance of b-jet identification in the ATLAS experiment. J** Instrum 11(04):P04008–P04008. https://doi.org/10.1088/1748-0221/11/04/p04008
65. ATLAS Collaboration (2015) Expected performance of the ATLAS b -tagging algorithms in Run-2. Technical report Geneva: CERN. https://cds.cern.ch/record/2037697
66. **ATLAS Collaboration (2017) Secondary vertex finding for jet flavour identification with the** ATLAS detector. Technical report Geneva: CERN. https://cds.cern.ch/record/2270366
67. **Billoir P, Fruhwirth R, Regler M (1985) Track element merging strategy and vertex fitting** in complex modular detectors. Nucl Instrum Meth A 241:115–131. https://doi.org/10.1016/0168-9002(85)90523-6
68. **Billoir P, Qian S (1992) Fast vertex fitting with a local parametrization of tracks. Nucl Instrum** Meth A 311:139–150. https://doi.org/10.1016/0168-9002(92)90859-3
69. **ATLAS Collaboration (2018) Topological b -hadron decay reconstruction and identification** of b - jets with the JetFitter package in the ATLAS experiment at the LHC. Technical report Geneva: CERN. https://cds.cern.ch/record/2645405
70. Piacquadio G, Weiser C (2008) A new inclusive secondary vertex algorithm for b-jet tagging in ATLAS. J Phys Conf Ser 119(3):032032. https://doi.org/10.1088/1742-6596/119/3/032032
71. **ATLAS Collaboration (2019) ATLAS b-jet identification performance and efficiency measure**ment with $t\bar{t}$ events in pp collisions at $\sqrt{s} = 13$ TeV. Eur Phys J C 79(11). ISSN: 1434-6052. https://doi.org/10.1140/epjc/s10052-019-7450-8
72. **ATLAS Collaboration (2018) Measurements of b-jet tagging efficiency with the ATLAS detec**tor using t t events at $\sqrt{s} = 13$ TeV. J High Energy Phys 2018(8). ISSN: 1029-8479. https://doi.org/10.1007/jhep08(2018)089
73. **ATLAS Collaboration (2015) Reconstruction, energy calibration, and identification of hadron**ically decaying tau leptons in the ATLAS experiment for Run-2 of the LHC. Technical report Geneva: CERN. https://cds.cern.ch/record/2064383
74. Aad G et al (2016) Reconstruction of hadronic decay products of tau leptons with the ATLAS experiment. Eur Phys J C 76(5). ISSN: 1434-6052. https://doi.org/10.1140/epjc/s10052-016-4110-0
75. ATLAS Collaboration (2018) Performance of missing transverse momentum reconstruction with the ATLAS detector using proton–proton collisions at $\sqrt{s} = 13$ TeV. Eur Phys J C 78(11). ISSN: 1434-6052. https://doi.org/10.1140/epjc/s10052-018-6288-9

Chapter 4
Search for the $VH(H \to c\bar{c})$ Decay

4.1 Introduction

Since the discovery of the Higgs boson in 2012 by the ATLAS and CMS Collaborations [1, 2], many of the Higgs production and decay modes have been observed, and the measured Higgs boson properties show consistency with Standard Model (SM) expectations. Interactions with third-generation fermions have been observed by the ATLAS [3–5] and CMS [6–8] Collaborations. Following this, evidence for interactions with second-generation fermions is beginning to emerge, starting with Higgs decays to muons, where CMS has recently reported evidence [9], and ATLAS has observed a 2σ excess over the background-only hypothesis [10]. Evidence for the Higgs coupling to other first and second-generation particles is still elusive. Direct searches for $H \to c\bar{c}$ decays [11, 12], $H \to e^+e^-$ decays [13, 14] and measurements of exclusive decays to mesons [15–18] are able to provide upper limits on the Higgs couplings to c-quarks, s-quarks and electrons, but show no direct evidence.

In the study of the Higgs coupling to first and second-generation fermions, the Higgs decay to charm-quarks is the most promising decay mode to investigate. With a branching ratio of 2.89% [19], $H \to c\bar{c}$ is one of the most common Higgs decay modes which has not been observed yet. Additionally, due to the smallness of the Higgs-charm Yukawa coupling, it is susceptible to significant modifications in various new physics scenarios [20–26].

Direct searches for $H \to c\bar{c}$ decays have set upper limits on the cross-section times branching fraction of this process. The ATLAS Collaboration has performed a search in the $ZH \to \ell\ell c\bar{c}$ decay channel, which set an upper limit of 110 times the signal strength predicted by the Standard Model (110 $\times$ SM, for an expected limit of 150 $\times$ SM) [11]. A recent search by the CMS Collaboration observed an upper limit of 37 $\times$ SM (for 70 $\times$ SM expected) [12].

This chapter provides an overview of a new search for Higgs boson decays to charm quarks with the ATLAS detector. The search is performed in the VH production mode, where the vector boson (W or Z) decays leptonically. The analysis

M. Mironova, *Search for Higgs Boson Decays to Charm Quarks with the ATLAS Experiment and Development of Novel Silicon Pixel Detectors*, Springer Theses,
https://doi.org/10.1007/978-3-031-36220-0_4

utilises the full dataset of pp collisions collected by the ATLAS detector during Run 2 of the LHC, corresponding to an integrated luminosity of 139 fb^{-1}, collected at a centre-of-mass energy of 13 TeV. The analysis is conducted in three channels, targeting different decay modes of the vector boson: $ZH \to \nu\nu c\bar{c}$ (*0-lepton* channel), $WH \to \ell\nu c\bar{c}$ (*1-lepton* channel) and $ZH \to \ell\ell c\bar{c}$ (*2-lepton* channel). This analysis expands on the previous search conducted by the ATLAS Collaboration by utilising a larger pp collision dataset, introducing the *0-lepton* and *1-lepton* channels, as well as several analysis improvements relating to the identification of c-quark jets and the background estimation.

An overview of this analysis is provided in Sect. 4.2. Section 4.3 introduces the data and simulated samples used in the analysis. Section 4.4 outlines the object and event selection and Sect. 4.5 provides details about the flavour tagging selection used. Following this, the experimental uncertainties are discussed in Sect. 4.6 and a detailed description of uncertainties related to the modelling of the signal and background processes can be found in Sect. 4.7. The methodology of the statistical analysis is covered in Sect. 4.8, and the results of the $VH(H \to c\bar{c})$ search are presented in Sect. 4.9. The result of a simple statistical combination with the $VH(H \to b\bar{b})$ measurement is described in Sect. 4.10. Finally, the expected sensitivity of this analysis is extrapolated to the High-Luminosity LHC in Sect. 4.11.

4.2 Analysis Overview

This search for Higgs decays into charm-quarks targets Higgs bosons produced in association with a vector boson, i.e. in the VH production mode. In pp collisions, the background from QCD events is overwhelming, especially for analyses searching for final states with quarks, such as $H \to c\bar{c}$. The QCD multi-jet background can be significantly reduced by selecting leptonic decays of the vector boson in the VH production mode, which are characterised by large E_T^{miss}, or high momentum leptons. Based on the decay of the vector boson, events are categorised into $ZH \to \nu\nu c\bar{c}$ (*0-lepton* channel), $WH \to \ell\nu c\bar{c}$ (*1-lepton* channel) and $ZH \to \ell\ell c\bar{c}$ (*2-lepton* channel). A candidate $VH, H \to c\bar{c}$ event is shown in Fig. 4.1. This is a *2-lepton* event with a $ZH \to \mu\mu c\bar{c}$ decay, and hits of the muons in the muon spectrometer and the reconstructed tracks can be seen. The $c\bar{c}$ pair is inferred from the two reconstructed jets. In this analysis, c-jet tagging is used to identify $H \to c\bar{c}$ candidate events and suppress backgrounds. The analysis regions are defined using c-tagging and selections on the kinematic variables. *Signal regions* (SRs) have been optimised to maximise the sensitivity to the $VH(H \to c\bar{c})$ process, and *control regions* (CRs) have been defined to estimate the background processes using data events. The invariant mass of the two highest momentum jets in the event (m_{cc}) is used as the discriminating variable. A *binned likelihood fit* to the m_{cc} distributions is performed in order to extract a measurement of the *signal strength* of the $VH(H \to c\bar{c})$ process. The signal strength is defined as the cross-section times branching ratio of the process, divided by its SM prediction. In addition to the $VH(H \to c\bar{c})$ decay,

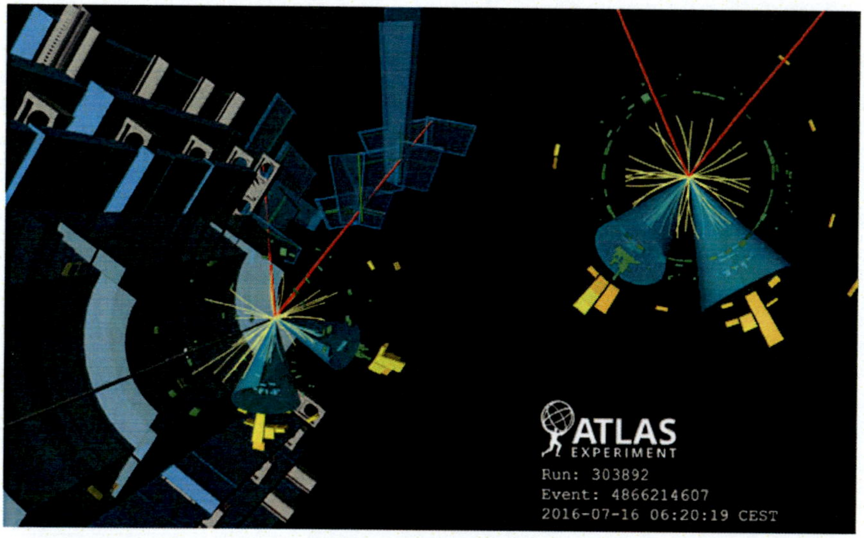

Fig. 4.1 Candidate event for the process $ZH \rightarrow \mu\mu c\bar{c}$ (Run 309892, Event 4866214607). Two muons, which have a di-lepton invariant mass of 92 GeV, are shown as red tracks producing hits (green) in the endcap muon chambers. The reconstructed Z boson has a transverse momentum of 150 GeV. The Higgs boson candidate is reconstructed from two charm-tagged $R = 0.4$ jets (blue cones), which have associated energy deposits in both the electromagnetic (green) and hadronic (yellow) calorimeters. The leading and sub-leading jets have transverse momenta of 123 GeV and 71 GeV respectively. The Higgs boson candidate has a reconstructed invariant mass of 123 GeV

the diboson processes $VZ(Z \rightarrow c\bar{c})$ and $VW(W \rightarrow cq)$ $(q = s, d)$ are measured as a cross-check of the analysis strategy and a validation of the c-tagging procedure. The measurement of the $VH(H \rightarrow c\bar{c})$ process can also be used to extract a direct constraint on the Higgs-charm Yukawa coupling. Finally, due to the similarity of the two processes, a simultaneous measurement of $VH(H \rightarrow c\bar{c})$ and $VH(H \rightarrow b\bar{b})$ can be conducted, by considering the analysis regions of this analysis and combining them with the dedicated $VH(H \rightarrow b\bar{b})$ measurement. Using this combination, it is possible to measure the Higgs-charm and Higgs-bottom Yukawa coupling simultaneously.

4.3 Data and Simulated Samples

4.3.1 Data

The analysis utilises a dataset collected by the ATLAS experiment during Run 2 of the LHC, corresponding to a total luminosity of 139 fb^{-1} at a centre-of-mass energy of 13 TeV. The data was collected using $E_{\mathrm{T}}^{\mathrm{miss}}$ triggers in the *0-lepton* and

Table 4.1 Signal and background processes and their corresponding MC generators used in this analysis. The acronyms ME and PS stand for matrix element and parton shower, respectively. The cross-section order refers to the order of the cross-section calculation used for process normalisation in QCD, unless otherwise stated, with ((N)N)LO and ((N)N)LL standing for ((next-to-)next-to-) leading order and ((next-to-)next-to-) leading log, respectively

Process	ME generator	ME PDF	PS and hadronisation	Tune	Cross-section order
$qq \to VH$ $(H \to c\bar{c}/b\bar{b})$	POWHEG- BOX v2 [27] + GOSAM [36] + MINLO [37, 38]	NNPDF3.0NLO [28]	PYTHIA 8.212 [29]	AZNLO [30]	NNLO(QCD) +NLO(EW) [31–35]
$gg \to ZH$ $(H \to c\bar{c}/b\bar{b})$	POWHEG- BOX v2	NNPDF3.0NLO	PYTHIA 8.212	AZNLO	NLO+NLL [39, 40]
$t\bar{t}$	POWHEG- BOX v2 [41]	NNPDF3.0NLO	PYTHIA 8.230	A14 [42]	NNLO +NNLL [43–49]
t/s-channel single top	POWHEG- BOX v2 [50]	NNPDF3.0NLO	PYTHIA 8.230	A14	NLO [51, 52]
Wt-channel single top	POWHEG- BOX v2 [53]	NNPDF3.0NLO	PYTHIA 8.230	A14	Approx. NNLO [54, 55]
V+jets	SHERPA 2.2.1 [56–58]	NNPDF3.0NNLO [28]	SHERPA 2.2.1	Default	NNLO [59]
$qq \to VV$	SHERPA 2.2.1	NNPDF3.0NNLO	SHERPA 2.2.1	Default	NLO
$gg \to VV$	SHERPA 2.2.2	NNPDF3.0NNLO	SHERPA 2.2.2	Default	NLO

1-lepton channels and single-lepton triggers in the *1-lepton* and *2-lepton* channels. Data is required to be of good quality while the relevant detector components are in operation.

4.3.2 Simulated Samples

The signal and background processes in the analysis are modelled using Monte Carlo (MC) simulation, apart from a data-driven estimate of the QCD multi-jet background. The different relevant signal and background processes and MC samples are summarised in Table 4.1 and described in more detail below.

Signal

The $VH(H \to c\bar{c})$ signals considered in this analysis are the three main qq production modes, as illustrated in Fig. 4.2. In the case of the ZH process, the production can also be initiated by gluons at next-to-leading order (NLO), via a top-quark loop. Both the $q\bar{q} \to ZH$ and $gg \to ZH$ processes are taken into account. For the $VH \to c\bar{c}$ signal, the matrix element calculation is performed using the POWHEG-

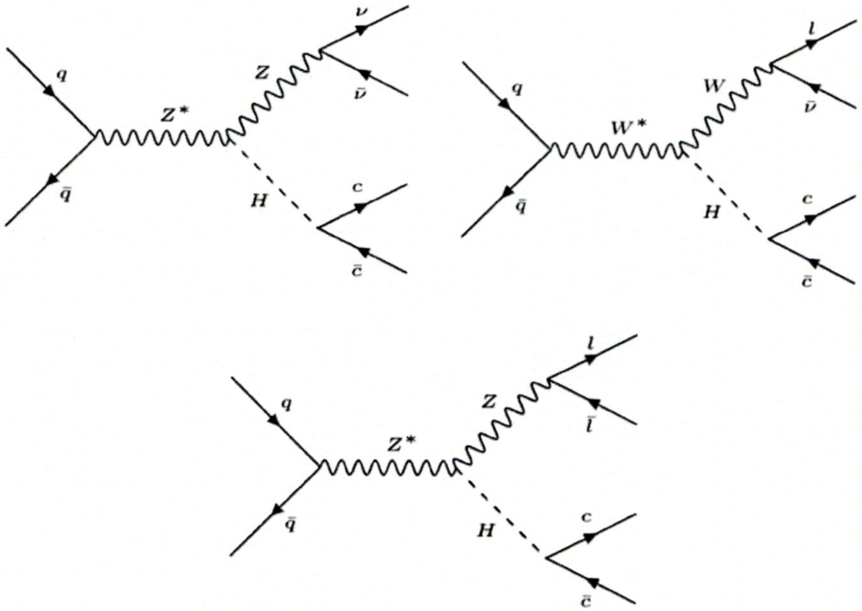

Fig. 4.2 Leading order Feynman diagrams of the $VH(H \to c\bar{c})$ signal processes

BOX v2 generator [27, 60], using the NNPDF3.0 PDF set [28]. The parton shower and hadronisation are simulated using PYTHIA 8.230 [29]. The samples are generated assuming a Higgs mass of $m_H = 125$ GeV and a SM branching ratio of 0.02891 [61]. For the assessment of systematic modelling uncertainties, additional samples are generated using POWHEG + HERWIG 7.

V+jets Background

The production of a W or Z boson associated with additional jets (i.e. V+jets) represents the largest background in the analysis. Figure 4.3 shows two examples of V+jets production modes that may pass the event selection. On the left, the additional jets are produced through gluon splitting, resulting in two jets of the same flavour, whereas on the right, different flavours of jets are possible.

The V+jets processes are simulated with SHERPA 2.2.1 [56–58] interfaced with NNPDFs [28] for both the ME calculation and the parton shower simulation. Since this analysis targets processes containing heavy flavour quarks, the V+jets samples are generated using different filters to select the flavour composition of the jets associated with the vector boson.

Alternative MC samples are also considered to validate the modelling of the generated SHERPA 2.2.1 samples and for the derivation of systematic uncertainties.

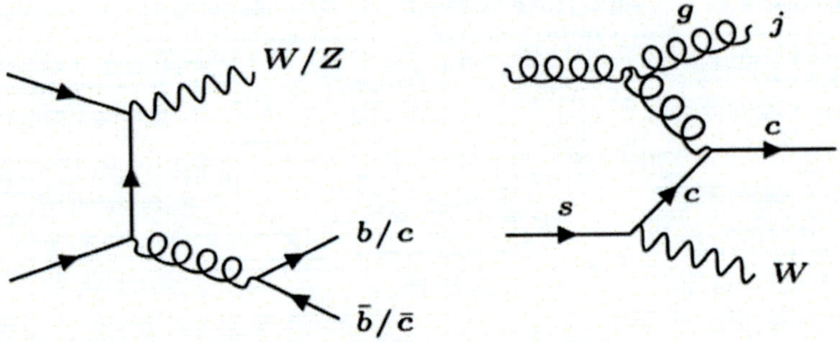

Fig. 4.3 Examples of Feynman diagrams of V+jets processes included in the analysis

Fig. 4.4 Feynman diagram of the $t\bar{t}$ production and decay

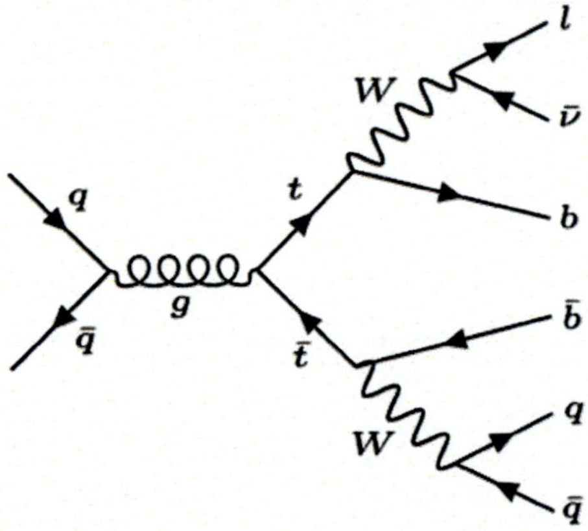

The alternative samples for the V+jets processes are generated using MADGRAPH 5 [62], interfaced to PYTHIA 8 for the modelling of the parton shower.

Top-Quark Background

The second-largest background in the analysis originates from top quark pair production ($t\bar{t}$). The production and decay of a $t\bar{t}$ event is shown in Fig. 4.4. The $t\bar{t}$ background enters the event selection in different ways in the various analysis channels. In the *0-lepton* and *1-lepton* channel, the reconstructed vector boson system originates from a single W boson from the $t\bar{t}$ decay. In contrast, in the *2-lepton* channel, the di-lepton system is reconstructed using leptons from the two different top quarks. The $t\bar{t}$ **background is significant in the** *0-lepton* and *1-lepton* channels and small in the *2-lepton* channel.

The nominal MC sample for the $t\bar{t}$ process is generated using POWHEG + PYTHIA 8, which utilises the POWHEG NLO matrix element (ME) generator [41] interfaced to PYTHIA 8.210 [29] for the parton shower and hadronisation. To assess the

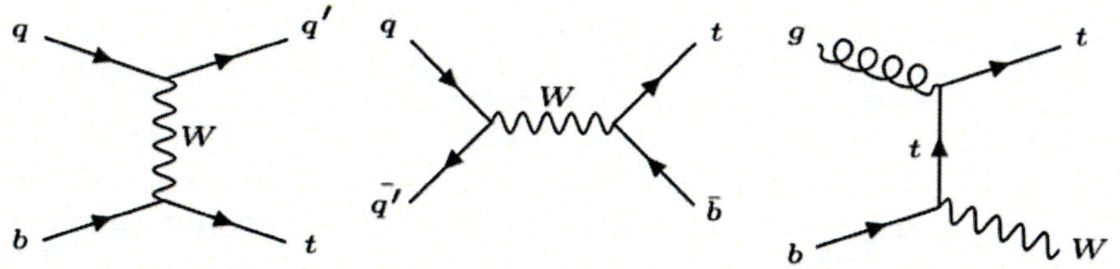

Fig. 4.5 Feynman diagrams of single top production in the t-channel (left), s-channel (center) and Wt-channel (right)

modelling uncertainties, different alternative samples for the $t\bar{t}$ processes are considered, targeting different sources of uncertainties:

- **POWHEG+HERWIG 7—Parton Shower**: HERWIG 7.0 [63] is used for simulating the parton shower, providing a handle on uncertainties related to systematically varying the parton shower model (HERWIG 7 vs. PYTHIA 8) with respect to the nominal prediction.
- **MADGRAPH 5_aMC@NLO+PYTHIA 8.2—Matrix Element**: In this alternative sample, MADGRAPH 5_aMC@NLO [62] is used for the hard scattering generation at NLO precision, while PYTHIA 8.2 is kept for the simulation of the parton shower. This sample allows to systematically vary the matrix element prediction with respect to the nominal setup.
- **POWHEG + PYTHIA 8—Initial/Final State Radiation** Three different samples are produced to study the impact of initial and final state radiation:
  - Initial state radiation (ISR) *RadLo*: The renormalisation scale μ_R and factorisation scale μ_F are doubled, while the other POWHEG +PYTHIA 8 parameters are kept at their nominal values.
  - Initial state radiation (ISR) *RadHi*: The renormalisation scale μ_R and factorisation scale μ_F are halved while the other POWHEG +PYTHIA 8 parameters are kept at their nominal values, except `hdamp`, which is doubled from its nominal value to $3.0 \times m_{top}$.
  - Final state radiation (FSR) variations: The renormalisation scale $\mu_{R,FSR}$ used in the PYTHIA 8 parton shower code is varied. It is either halved (down variation) or doubled (up variation).

A minor background in this analysis comes from single top-quark production. These events can be produced in the s-, t- and Wt-channel, as illustrated in Fig. 4.5. The dominant contributions comes from the Wt production mode. POWHEG + PYTHIA 8 is also used to simulate these processes. As for the $t\bar{t}$ background, the same set of alternative samples is available for the Wt process. An additional alternative sample is used to quantify the impact of using Diagram Removal (DR, nominal) or Diagram Subtraction (DS) to treat the overlap of $t\bar{t}$ and Wt events.

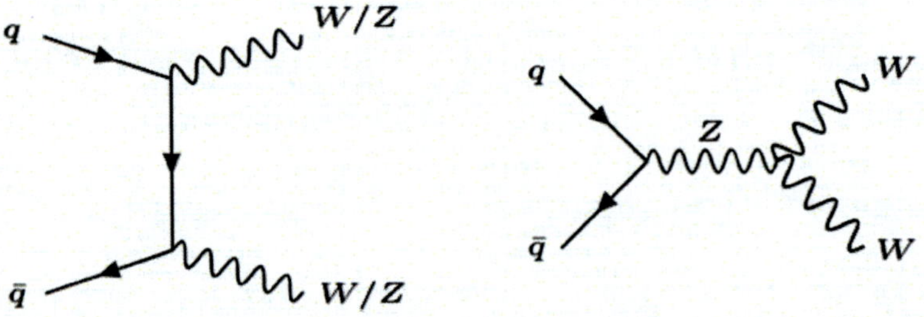

Fig. 4.6 Feynman diagram of the diboson production in the t-channel (left) and s-channel (right)

Diboson Signal and Background

Different diboson processes can enter the analysis (WW, WZ and ZZ) and examples of the diboson processes are shown in Fig. 4.6. In addition to the quark-induced diagrams, gluons can initiate diboson events via quark loops. Diboson events can leave a similar signature as $VH(H \to c\bar{c})$ in the detector when one of the vector bosons decays leptonically and the other hadronically and these events are used as a cross-check signal in the analysis. MC events are generated for these signatures using SHERPA 2.2.1 for the $qqVV$ processes and SHERPA 2.2.2 for the $ggVV$ processes. Additional samples are generated using POWHEG + PYTHIA 8 to assess the systematic modelling uncertainties on the diboson processes.

$VH(H \to b\bar{b})$ Background

The $VH(H \to b\bar{b})$ process has the same signature and similar kinematics as the $VH(H \to c\bar{c})$ process, except for the Higgs decay into two b-quarks. The process is simulated using POWHEG + PYTHIA 8, and alternative samples are generated using POWHEG + HERWIG 7.

QCD Multi-jet Background

The QCD multi-jet background is negligible in the *0-lepton* and *2-lepton* channel, thus no dedicated QCD multi-jet background contribution is considered. In the *1-lepton* channel, the QCD background is a minor background and is derived in a data-driven approach using a multi-jet enriched region.

4.4 Selection and Categorisation

4.4.1 Object Selection

This section outlines the definitions of the reconstructed physics objects used in the analysis.

Primary Vertices

Events that enter the analysis are required to have at least one reconstructed primary vertex, based on the primary vertex reconstruction using tracks in the ID [64].

Electrons

Electrons are reconstructed from ID tracks and matching energy clusters in the electromagnetic calorimeter [65]. All electrons are required to have $p_T > 7$ GeV and the electron cluster should be in the range $|\eta| < 2.47$. All electrons entering the analysis must pass the *loose* identification criterion. They must also be isolated both in the ID and in the calorimeter based on p_T-dependent criteria. In the *1-lepton* and *2-lepton* channels, the electrons are required to have $p_T > 27$ GeV. In the *1-lepton* channel, more stringent isolation and identification criteria are in place (using the *tight* selection criteria), to suppress QCD multi-jet background.

Muons

Muons are reconstructed from tracks in the ID and corresponding tracks in the muon spectrometer [66]. All muons are required to have $p_T > 7$ GeV and $|\eta| < 2.5$, satisfy the *loose* identification criteria, and p_T-dependent isolation criteria in the ID. Additionally, in the *2-lepton* channel muons are required to have $p_T > 27$ GeV. In the *1-lepton* channel more stringent criteria are applied, requiring $p_T > 25$ GeV and a tighter isolation requirement.

τ-Leptons

Reconstructed τ-leptons are not directly used in the analysis, but they are used in the calculation of $E_{\mathrm{T}}^{\mathrm{miss}}$ and to avoid double-counting reconstructed τ-leptons as other objects, e.g. jets. Hadronically decaying τ-leptons [67, 68] are identified with a *medium* quality criterion [68]. They are required to have $p_T > 20$ GeV and $|\eta| < 2.5$, excluding the transition region between the barrel and endcap of the electromagnetic calorimeters of $1.37 < |\eta| < 1.52$, and are required to have 1 or 3 tracks.

Missing Transverse Energy

The missing transverse energy, $E_{\mathrm{T}}^{\mathrm{miss}}$, plays an important role in the *0-lepton* and *1-lepton* channels. It is reconstructed using the negative vectorial sum of the transverse momenta of the physics objects and the soft term [69]. Missing momentum can also be reconstructed based on the ID tracks, in a quantity called $E_{\mathrm{T,trk}}^{\mathrm{miss}}$ [70]. This quantity can be used in cuts to suppress the non-collisional background in the analysis.

Jets

Jets are an integral part of the analysis. They are reconstructed by using topological calorimeter-cell clusters, which are calibrated using the EM scale [71–73]. The clusters are used as input to the anti-k_T algorithm [74, 75], where the jets are reconstructed using a radius parameter of $R = 0.4$. Jets are classified as either *central* or *forward*. Forward jets are required to have $2.5 < |\eta| < 4.5$ and $p_T > 30$ GeV. Central jets must have $p_T > 20$ GeV and $|\eta| < 2.5$. In addition, central jets with $p_T < 120$ GeV must originate from the primary vertex, as identified by the Jet Vertex Tagger (JVT) [76]. Central jets can be used in flavour-tagging and are subsequently used in the reconstruction of the Higgs boson candidate.

The jets are reconstructed at the *Global Sequential Calibration* (GSC) scale. From this, the jet momentum scale and resolution can be improved using jet energy corrections, which can improve the m_{cc} invariant mass resolution. Loss of reconstructed energy can be caused by semi-leptonic decays inside the jets, where muons or neutrinos are not properly accounted for, as they deposit little or no energy in the calorimeters. To improve the jet resolution, a *muon-in-jet* correction is applied. If any muons are found within a p_T-dependent cone around the jet axis, the four-momentum of the muon closest to the jet is added to the four-momentum of the jet [77].

To avoid double-counting of physics objects, an overlap removal procedure is applied for electrons, muons, hadronically decaying τ-leptons and jets.

Each jet in the simulated events is assigned a *truth-flavour* label, either *b*-, *c*-, τ-jets or light jets. *b*-, *c*- and τ-jets are identified in this order, depending on whether the jet contains a *b*-hadron, *c*-hadron or τ-lepton with $p_T > 5$ GeV within a cone of $R = 0.3$ of their axis. Jets not identified in this procedure are labelled as light jets.

4.4.2 Event Selection

This section describes the event selection, which is summarised in Table 4.2. First, the common event selection is discussed, followed by the specific selections in the *0-lepton*, *1-lepton* and *2-lepton* channels.

Table 4.2 Summary of the signal region event selection in the *0-lepton*, *1-lepton* and *2-lepton* channels. **Jet1** and **jet2** refer to the two signal jets and **H** refers to the **jet1–jet2** system

Common selections	
Central jets	≥ 2
Signal jet p_T	≥ 1 signal jet with $p_T > 45$ GeV
c-jets	One or two c-tagged signal jets
b-jets	No b-tagged non-signal jets
Jets	2, 3 (*0-* and *1-lepton*); 2, ≥ 3 (*2-lepton*)
p_T^V regions	75–150 GeV (*2-lepton*) >150 GeV
ΔR(**jet1**, **jet2**)	$75 < p_T{}^V < 150$ GeV: $\Delta R \leq 2.3$ $150 < p_T{}^V < 250$ GeV: $\Delta R \leq 1.6$ $p_T{}^V > 250$ GeV: $\Delta R \leq 1.2$
0-lepton	
Trigger	E_T^{miss}
Leptons	No *loose* leptons
E_T^{miss}	>150 GeV
p_T^{miss}	>30 GeV
H_T	>120 GeV (2-jets), >150 GeV (3-jets)
$\min(\|\Delta\phi(E_T^{miss}, \textbf{jet})\|)$	$>20°$ (2-jets), $>30°$ (3-jets)
$\|\Delta\phi(E_T^{miss}, \textbf{H})\|$	$>120°$
$\|\Delta\phi(\textbf{jet1}, \textbf{jet2})\|$	$<140°$
$\|\Delta\phi(E_T^{miss}, p_T^{miss})\|$	$<90°$
1-lepton	
Trigger	e sub-channel: single electron μ sub-channel: E_T^{miss}
Leptons	One *tight* lepton and no additional *loose* leptons
E_T^{miss}	>30 GeV (e sub-channel)
m_T^W	<120 GeV
2-lepton	
Trigger	Single lepton
Leptons	Exactly two *loose* leptons Same flavour, opposite charge for $\mu\mu$
$m_{\ell\ell}$	$81 < m_{\ell\ell} < 101$ GeV

The transverse momentum of the vector boson (p_T^V) is reconstructed in different ways in the different channels. In the *0-lepton* channel the p_T^V is defined as the E_T^{miss}. In the *1-lepton* channel, the p_T^V is determined as the vectorial sum of E_T^{miss} and the lepton transverse momentum. In the *2-lepton* channel, the vector boson can be fully reconstructed, and the p_T^V is defined as the transverse momentum of the *2-lepton* system.

Because the signal to background ratio increases with increasing p_T^V, only events with $p_T^V > 75$ GeV are included in the analysis. Events are categorised into bins of p_T^V and jet multiplicity to improve sensitivity. In the *0-lepton* and *1-lepton* channel, a single bin of $p_T^V > 150$ GeV is considered. In the *2-lepton* channel a low p_T^V category ($75 < p_T^V < 150$ GeV) and a high p_T^V category ($p_T^V > 150$ GeV) are used. Events are selected with at least two central jets and are categorised as 2- and 3-jet events. Events with more than 3-jets are rejected in the *0-lepton* and *1-lepton* channel to suppress the $t\bar{t}$ background. In the *2-lepton* channel, events with 3 or more jets are included in the 3-jet category. The central jets are tagged as containing either b- or c-hadrons using two discriminants resulting from multivariate tagging algorithms, MV2 and DL1 [78]. More details about the flavour tagging selection are provided in Sect. 4.5. Events with 1 or 2 c-tags on the two leading p_T jets are selected in the signal region, and the 1 and 2 c-tag events are considered in separate categories. All c-tagged jets are also required not to be b-tagged, and any additional jets in the event are also required not to be b-tagged. The Higgs boson candidate is reconstructed from the two jets with the highest p_T, and the invariant mass of the two jets (m_{cc}) is calculated and used as the discriminating variable in this analysis. To further suppress the background contamination in all channels, cuts on the ΔR between the two signal jets are applied. An optimisation was performed in different regions of p_T^V to maximise sensitivity while retaining approximately 80% of the signal. As a result, the ΔR between the two signal jets is required to be:

- $\Delta R < 2.3$ in events with $75 < p_T^V < 150$ GeV
- $\Delta R < 1.6$ in events with $150 < p_T^V < 250$ GeV
- $\Delta R < 1.2$ in events with $p_T^V > 250$ GeV

Selection Specific to the *0-lepton* Channel

Events are selected in the *0-lepton* channel using the lowest unprescaled $E_{\mathrm{T}}^{\mathrm{miss}}$ trigger during each data collection period and requiring no *loose* leptons in the event. The $E_{\mathrm{T}}^{\mathrm{miss}}$ thresholds varied between 70 GeV in 2015 and 110 GeV in 2018. The efficiency of the $E_{\mathrm{T}}^{\mathrm{miss}}$ trigger is measured in W+jets, Z+jets and $t\bar{t}$ events using single-muon triggered data. Correction factors, which depend on the measured efficiency, are applied to the simulated events, with values ranging from 0.95 at a $E_{\mathrm{T}}^{\mathrm{miss}}$ of 150 GeV, to a negligible deviation from unity above values of 200 GeV. The non-collisional background is removed by requiring $E_{\mathrm{T,\,trk}}^{\mathrm{miss}} > 30$ GeV, where $E_{\mathrm{T,\,trk}}^{\mathrm{miss}}$ is the sum of the transverse momenta of the tracks associated with the primary vertex of the event. The QCD multi-jet background is removed by applying several different selections, also referred to as the *anti-QCD cuts*:

- $|\Delta\Phi(\mathbf{E}_{\mathrm{T}}^{\mathrm{miss}}, \mathbf{E}_{\mathrm{T,\,trk}}^{\mathrm{miss}})| < 90°$
- $|\Delta\Phi(\mathbf{jet1}, \mathbf{jet2})| < 140°$
- $|\Delta\Phi(\mathbf{E}_{\mathrm{T}}^{\mathrm{miss}}, \mathbf{H})| > 120°$
- $\min[|\Delta\Phi(\mathbf{E}_{\mathrm{T}}^{\mathrm{miss}}, \text{pre-selected jets})|] > 20°$ for 2-jets, $> 30°$ for 3-jets.

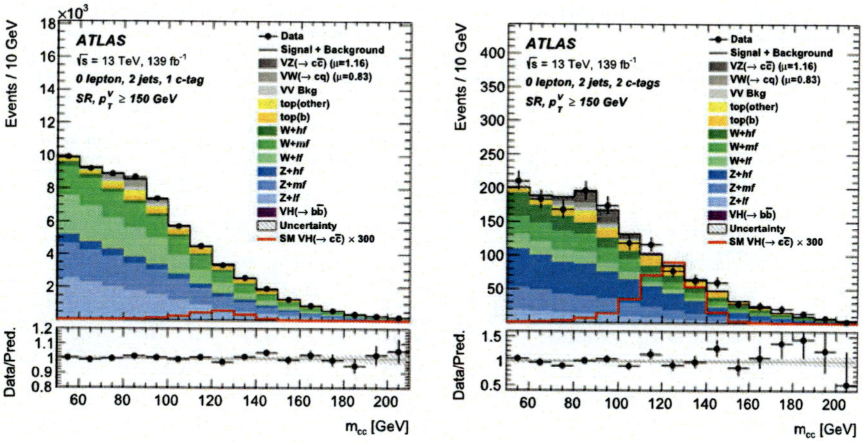

Fig. 4.7 m_{cc} distributions for selected signal regions in the *0-lepton* channel, with two jets and $p_T^V > 150$ GeV, for 1 c-tag (left) and 2 c-tags (right). The signal and backgrounds are scaled to their best-fit values. The total signal-plus-background prediction is shown by the solid black line. The $H \to c\bar{c}$ signal is also shown as an unfilled histogram scaled to 300 times the SM prediction. The post-fit uncertainty is shown as the hatched background

where Φ is the azimuthal angle and **jet1** and **jet2** are the two selected jets forming the Higgs candidate **H**, and $E_{\mathrm{T,\,trk}}^{\mathrm{miss}}$ is the track-based missing transverse momentum. After these cuts, the remaining multi-jet contribution in the *0-lepton* channel is less than 1%, which is negligible. The resulting m_{cc} distributions after the *0-lepton* event selection are shown in Fig. 4.7.

Selection Specific to the *1-lepton* Channel

The *1-lepton* channel has a different event selection, depending on whether the selected lepton is an electron or a muon. If the lepton is an electron (muon), it must have $p_T > 27$ (25) GeV and $|\eta| < 2.47$ (2.5). In the muon channel, the data is collected with the same $E_{\mathrm{T}}^{\mathrm{miss}}$ triggers as used in the *0-lepton* channel. For the online $E_{\mathrm{T}}^{\mathrm{miss}}$ calculation, muons are not considered, so effectively, these triggers select high p_T^V muon events and are more efficient than single-muon triggers. The lowest unprescaled single electron triggers in each collection period are used in the electron channel. The trigger threshold ranges between 20 and 140 GeV, depending on the electron ID requirement. There can be a significant background from events where jets are misidentified as electrons at low $E_{\mathrm{T}}^{\mathrm{miss}}$. These events can be suppressed by requiring $E_{\mathrm{T}}^{\mathrm{miss}} > 30$ GeV, in the electron-channel only. Finally, in both the electron and muon channel, the transverse mass of the reconstructed W boson,[1] m_{T}^{W}, is

[1] $m_{\mathrm{T}}^{W} = \sqrt{2p_T^{\ell}p_T^{\nu}(1 - \cos(\phi^{\ell} - \phi^{\nu}))}$, where $E_{\mathrm{T}}^{\mathrm{miss}}$ is used as an approximation for p_{T}^{ν}.

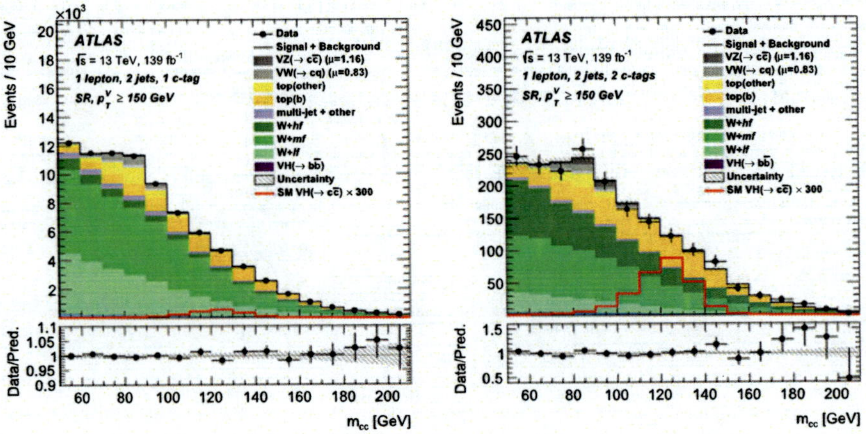

Fig. 4.8 m_{cc} distributions for selected signal regions in the *1-lepton* channel, with two jets and $p_T^V > 150$ GeV, for 1 c-tag (left) and 2 c-tags (right). The signal and backgrounds are scaled to their best-fit values. The total signal-plus-background prediction is shown by the solid black line and includes the $H \to c\bar{c}$ signal. The $H \to c\bar{c}$ signal is also shown as an unfilled histogram scaled to 300 times the SM prediction. The post-fit uncertainty is shown as the hatched background

required to be less than 120 GeV. The resulting m_{cc} distributions after the *1-lepton* event selection are shown in Fig. 4.8.

As the QCD multi-jet background in the *1-lepton* channel is non-negligible, it is accounted for in the analysis with a dedicated template. The multi-jet background is estimated in a data-driven approach by deriving the estimated number of multi-jet events and the m_{cc} shape in a dedicated control region. This control region is defined by inverting the tight isolation requirements on the leptons and subtracting all simulated backgrounds. Different templates are used in the electron and muon channel and in the different jet multiplicity bins.

Selection Specific to the *2-lepton* Channel

In the *2-lepton* channel, events with two *loose* leptons are selected, considering the same-flavour combinations of ee and $\mu\mu$. The lowest single-electron and single-muon triggers are used, and one of the selected leptons is required to have fired the trigger. In the di-muon case, the leptons must also have opposite charge. In the di-electron case, no such requirement is made as the electron charge misidentification rate is higher. At least one of the leptons must have a transverse momentum of $p_T > 27$ GeV to be consistent with the trigger selection. Finally, the invariant mass of the two leptons is required to be compatible with the Z-boson mass, i.e. $81 < m_{\ell\ell} < 101$ GeV. The resulting m_{cc} distributions after the *2-lepton* event selection are shown in Fig. 4.9.

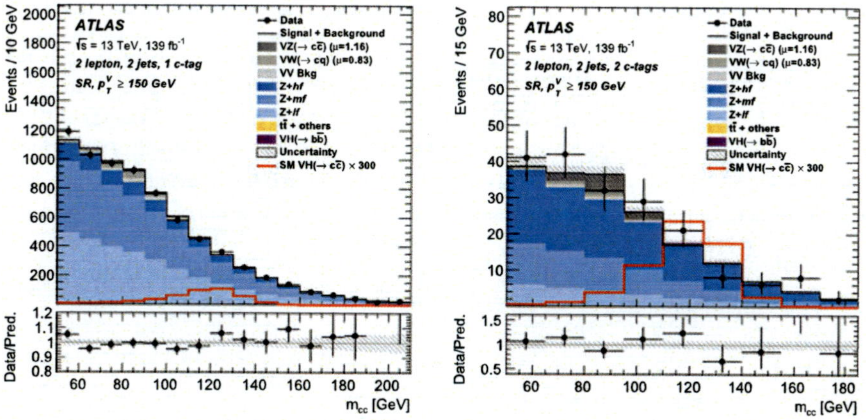

Fig. 4.9 m_{cc} distributions for selected signal regions in the *2-lepton* channel, with two jets and $p_T^V > 150$ GeV, for 1 c-tag (left) and 2 c-tags (right). The signal and backgrounds are scaled to their best-fit values. The total signal-plus-background prediction is shown by the solid black line and includes the $H \to c\bar{c}$ signal. The $H \to c\bar{c}$ signal is also shown as an unfilled histogram scaled to 300 times the SM prediction. The post-fit uncertainty is shown as the hatched background

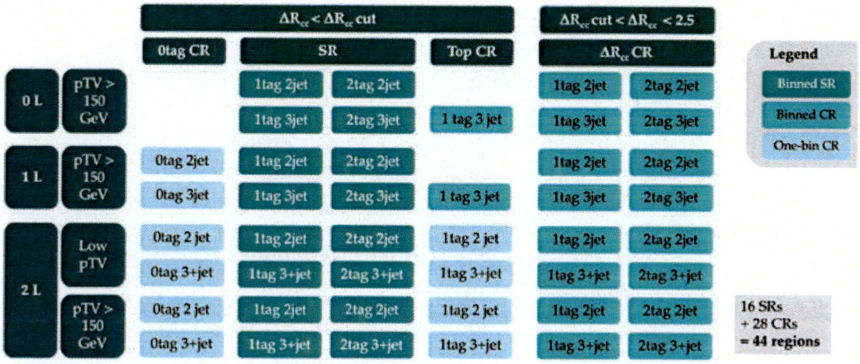

Fig. 4.10 Illustration of the signal and control regions which enter the $VH(H \to c\bar{c})$ analysis

4.4.3 Background Control Regions

The $VH(H \to c\bar{c})$ analysis is characterised by a complex background composition and a tiny signal. To ensure excellent estimation of all backgrounds, 28 *control regions* (CRs) are introduced, in addition to the 16 *signal regions* (SRs). The CRs are defined by modifying the event selection, such that a region is pure in a particular background. All signal and control regions are summarised in Fig. 4.10, and the control regions are described in detail below.

0 c-Tag Control Region

The 0 c-tag control regions follow the signal region event selection, except for a requirement that neither of the two leading jets are c-tagged. This results in a region enriched in light-flavour jets. In particular, this control region is introduced in the *1-lepton* channel to constrain the W+jets background containing light-flavour jets (W+light). In the *2-lepton* channel, the 0 c-tag control region constrains the Z+light background. One control region is added per jet multiplicity and p_T^V bin. The 0 c-tag control region is only used to determine the total number of V+light events, but not the m_{cc} shape.

Top Control Region

The top quark ($t\bar{t}$ and single top) background in the *0-lepton* and *1-lepton* channel is significant. In the *2-lepton* channel, the $t\bar{t}$ background is small, but non-negligible, particularly in the low p_T^V region. As a consequence, dedicated top control regions are introduced in all three channels.

In the *0-lepton* and *1-lepton* channel, a top control region is added in the 3-jet category by requiring the third jet in the event to be b-tagged using the 70% `MV2c10` working point (i.e. inverting the requirement for the signal region). This selection produces a region pure in $t\bar{t}$ and single top backgrounds. Due to limited statistics in the 2 c-tag case, only the 1 c-tag category is used, and one control region is introduced for both the *0-lepton* and *1-lepton* channels each. This control region is used to constrain the top quark background normalisation and m_{cc} shape. The resulting m_{cc} distributions in the *0-lepton* and *1-lepton* top control regions are shown in Fig. 4.11.

In the *2-lepton* channel, the top background originates mainly from dileptonic $t\bar{t}$ events, where both of the W-bosons in the $t\bar{t}$ events decay leptonically. A control region for this background is designed by requiring the two leptons in the event to have opposite flavour, i.e. $e\mu$ or μe. This removes signal events and other backgrounds, which are expected to have same flavour leptons from the Z-boson decay. Only the control region with 1 c-tag is used for simplicity, and one control region is added for each p_T^V and jet multiplicity category in the *2-lepton* channel. This control region is only used to determine the overall yield of $t\bar{t}$ events.

High ΔR Control Region

The high ΔR control region is designed to constrain the V+jets background normalisation and m_{cc} shape. It is selected by inverting the signal region ΔR cuts and introducing an upper cut-off at $\Delta R = 2.5$. This value was chosen, as the ΔR distributions are flat but then start falling above $\Delta R = 2.5$ in the *0-lepton* channel, due to the anti-QCD cuts, and statistics become limited. In summary, the control regions are selected to have:

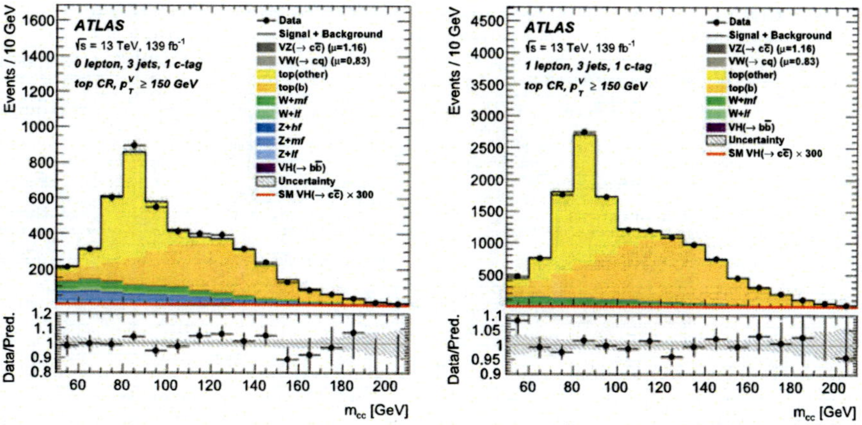

Fig. 4.11 m_{cc} distributions for the top control regions in the *0-lepton* (left) and *1-lepton* (right) channel. The signal and backgrounds are scaled to their best-fit values

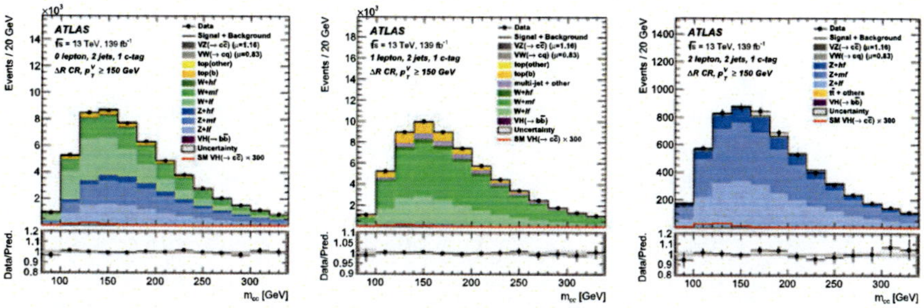

Fig. 4.12 m_{cc} distributions for the high ΔR control regions in the *0-lepton* (left), *1-lepton* (center) and *2-lepton* (right) channel. The signal and backgrounds are scaled to their best-fit values

- $2.3 < \Delta R < 2.5$ for events with $75 < p_T^V < 150$ GeV
- $1.6 < \Delta R < 2.5$ for events with $150 < p_T^V < 250$ GeV
- $1.2 < \Delta R < 2.5$ for events with $p_T^V > 250$ GeV

One control region is introduced for each signal region in all of the channels, resulting in 16 high ΔR control regions. Examples of the m_{cc} distributions in the ΔR CR for all three channels are shown in Fig. 4.12.

4.5 Flavour Tagging Selection

One of the most critical aspects of the analysis is the identification of c-jet using a c-tagger. Conceptually, the c-tagger is a multivariate algorithm that calculates the probability of a jet being a c-jet, or another flavour jet, based its properties. In addition

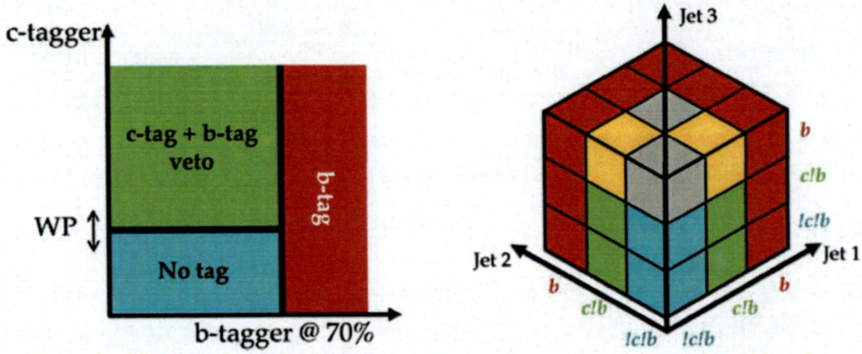

Fig. 4.13 Illustration of the two-dimensional c-tagging + b-veto working point (left). Illustration of flavour tagging categorisation in $VH(H \to c\bar{c})$ analysis events with three jets (right). The label b stands for a b-tagged jet, $c!b$ means c-tagged but not b-tagged, and $!c!b$ means not c-tagged and not b-tagged

to identifying c-jets, the secondary purpose of the c-tagger is to reject events in the $VH(H \to c\bar{c})$ analysis, which are already considered in the $VH(H \to b\bar{b})$ analysis, to allow for a combination of the two analyses at a later stage. Consequently, this analysis does not utilise flavour tagging in a binary scenario (*tagged or not tagged*), but considers three different cases, as illustrated in Fig. 4.13 (left): *c-tagged and not b-tagged, b-tagged* or *not c-tagged or b-tagged.* To ensure no overlap with the $VH(H \to b\bar{b})$ analysis, it is then possible to reject events with exactly two b-tags. A c-tagger for these purposes did not readily exist within the ATLAS Collaboration and was developed specifically for this analysis. The flavour tagging proceeds sequentially: first, it is checked whether a jet satisfies the b-tagging requirement, and if not, the c-tagging requirement is applied.

The c-tagging requirement is only applied to two leading jets in the analysis, and events with 1 or 2 c-tags are selected in the signal regions. Events with three or more jets can also enter the analysis, providing further possibilities where b-tagged jets can appear and require additional consideration. The $VH(H \to c\bar{c})$ flavour tagging approach is illustrated in Fig. 4.13 (right). The three axes indicate the three possible jets in the events, and the events are categorised by the flavour tagger as follows:

- If at least one of the two leading jets is c-tagged and not b-tagged, and in addition, the third jet in the event is not b-tagged, the event enters the $VH(H \to c\bar{c})$ signal regions, indicated as the green squares.
- If neither of the leading jets is c-tagged, and the third jet is not b-tagged, the event enters the 0 c-tag control region in the $VH(H \to c\bar{c})$ analysis, indicated by the blue squares.
- If the two leading jets are b-tagged, the event is discarded from the $VH(H \to c\bar{c})$ analysis and is likely to enter the $VH(H \to b\bar{b})$ analysis, indicated by the red squares.

- If the third jet in the event is b-tagged, and one of the leading jets is c-tagged, the event can enter the top control region of the $VH(H \to c\bar{c})$ analysis, indicated by the yellow squares.

The algorithm used for the c-tagging in this analysis combines the `DL1` and `MV2c10` discriminants [78] described in more detail in Sect. 3.3.6. As input variables, both algorithms use the log-likelihood ratios calculated by `IP2D` and `IP3D`, secondary vertex information from `SV1`, information about the relationship between the primary and secondary vertices from `JetFitter`, and the p_T and η of the jet. In `DL1`, this information is used in a Deep Neural Network, which then returns the probability of a jet being a b-jet (p_b), c-jet (p_c) or light-jet (p_l), while the sum of probabilities has to satisfy $p_b + p_c + p_l = 1$. In the case of `MV2c10`, a BDT is trained and returns the likelihood of a jet being a b-jet.

Most commonly, the output of the `DL1` algorithm is used for b-tagging, as described in Sect. 3.3.6, but it can also be used to define a one-dimensional c-tagging discriminant:

$$DL_1 = \ln \frac{p_c}{f p_b + (1-f) p_u} \tag{4.1}$$

Here, f corresponds to a weight attributed to the probabilities p_b and p_u. Both the value of DL_1 and f can be varied to change the working point of the c-tagger. A large value of f means better discrimination of c-jets against b-jets, while a low value of f means better discrimination against light-jets. In addition to the `DL1` c-tagger, a b-tagger based on the `MV2c10` discriminant is used to define the b-tag veto. Here the same working point as in the $VH(H \to b\bar{b})$ analysis is used, corresponding to an efficiency of 70% for correctly tagging b-jets [77].

The c-tagger and b-veto are applied together on the two leading jets. Thus, dedicated flavour tagging calibrations are needed for this combined working point to account for correlations between the c-tagger and the b-veto. On the additional jets, where the b-tagging is used standalone, the standard flavour tagging calibrations for the 70% `MV2c10` working point are applied.

4.5.1 Working Point Optimisation and Calibration

In the definition of the `DL1` discriminant, it is possible to vary the DL_1 cut and the fraction f to find an optimal configuration for the $VH(H \to c\bar{c})$ analysis. Such an optimisation was performed using events that pass the signal region selection in the three lepton channels. The DL_1 and f parameters are scanned in a grid while assuming the default b-tag veto at a 70% working point. In the *0-lepton* and *1-lepton* channels, the expected upper limit on the $VH(H \to c\bar{c})$ signal strength at 95% confidence level is used as a figure of merit. In the *2-lepton* channel, the significance $z = \frac{s}{\sqrt{s+b}}$ is maximised instead, where s and b corresponds to the number of signal and background events respectively.

Table 4.3 **Efficiencies of the c-tagging plus b-tag veto working point chosen by the optimisation study**

Requirement	Efficiency			
	c-jets	b-jets	light-jets	τ-jets
$DL_1 \geq 1.3$ and f $= 0.08$ and $MV2c10 \leq 0.83$	27%	8%	1.6%	25%

The ideal selection is found to be $DL_1 > 1.3$ with a weight value of $f = 0.08$ and is consistent for all three channels. The low value of $f = 0.08$ indicates that the signal sensitivity improves with better rejection against light-jets. This can be explained by the fact that many b-jet events are already vetoed, and the analysis uses a 1 c-tag region, where light-jets can play a significant role. The performance of this c-tagging working point on a sample of $t\bar{t}$ events is summarised in Table 4.3. The c-tagging efficiency for c-jets is 27%, illustrating the difficulty of distinguishing c-jets from b- and light-flavour jets. Notably, the c-tagging efficiency for τ-jets is 25%, due to the similar signature of c- and τ-jets in the detector.

To use the c-tagger in the analysis, scale factors (*SFs*) for the b-, c- and light-flavour jets have to be derived to ensure that the c-tagging efficiency is the same in data and the simulated samples. These flavour tagging calibration scale factors are derived as:

$$SF = \frac{\epsilon_{\text{Data}}}{\epsilon_{\text{MC}}} \tag{4.2}$$

where ϵ_{Data} and ϵ_{MC} correspond to the c-tagging efficiency of the flavour tagging algorithm in data and simulation respectively. Scale factors are derived in a data-driven study for each jet flavour individually. Before the scale factors can be used, they are smoothed using a local polynomial kernel estimator with a smoothing parameter of 0.4 to ensure smoothly varying SFs in the analysis and avoid introducing distortions in the kinematic variables [79].

Simulated samples with different parton shower models are used in the analysis, so differences in flavour tagging performance must be accounted for. The nominal calibration scale factors are derived using PYTHIA 8 samples. However, SHERPA 2.2.1 is also used to generate some simulation samples. To account for the differences, additional MC to MC scale factors are derived and can be multiplied to the nominal scale factors according to:

$$SF = \frac{\epsilon_{\text{Data}}}{\epsilon_{\text{MC,Py8}}} \times \frac{\epsilon_{\text{MC,Py8}}}{\epsilon_{\text{MC,Alternative}}}. \tag{4.3}$$

Furthermore, scale factors for jets failing the c-tagging requirements are also needed (*inefficiency scale factors*). These are derived from the efficiency scale factors as follows:

$$SF_{\text{inefficiency}} = \frac{1 - \epsilon_{\text{Data}}}{1 - \epsilon_{\text{MC}}} = \frac{1 - \epsilon_{\text{MC}} \times SF}{1 - \epsilon_{\text{MC}}}. \tag{4.4}$$

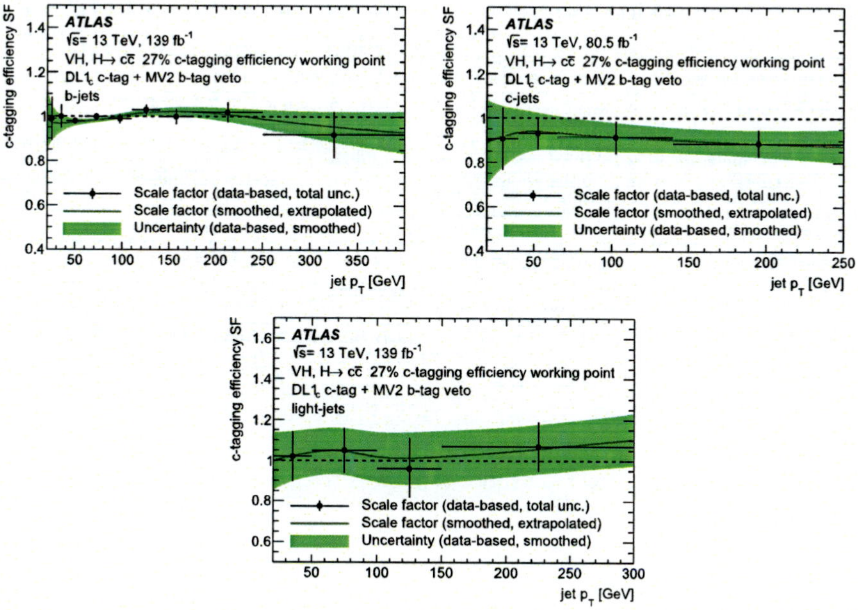

Fig. 4.14 Calibrated flavour tagging scale factors for c-jets, b-jets and light-jets for the c-tagging plus b-veto working point used in the $VH(H \rightarrow c\bar{c})$ analysis, using the `DL1` algorithm as a c-tagger and the `MV2c10` algorithm as a b-tagger

In summary, the following quantities are derived in dedicated calibration studies for this analysis:

- Calibration scale factors for b-, c-, τ- and light-jets.
- Finely binned flavour tagging efficiency maps, used for the computation of inefficiency scale factors and in truth-tagging (see Sect. 4.5.2).
- Coarsely binned flavour tagging efficiency maps used to compute the MC/MC scale factors.

The individual procedures to determine the flavour tagging scale factors are described in the following, and the resulting smoothed scale factors as a function of jet p_T are shown in Fig. 4.14.

b-jet SF calibrations The scale factors for the c-tagging efficiency of b-jets are derived in $t\bar{t}$ events in which both of the W bosons in the $t\bar{t}$ process decay leptonically [80]. Such a selection results in a sample pure in b-jets from the initial decay of the top quarks. Events are selected based on a BDT, which utilises different kinematic variables. PDFs of the tagging discriminant are built in several regions in the 2D space of p_T of the leading and subleading jet, where the bin edges are defined as shown in Fig. 4.14. These PDFs are then used to perform a likelihood fit to extract the c-tagging efficiency in the data sample. For $p_T^V < 250$ GeV, the uncertainties on

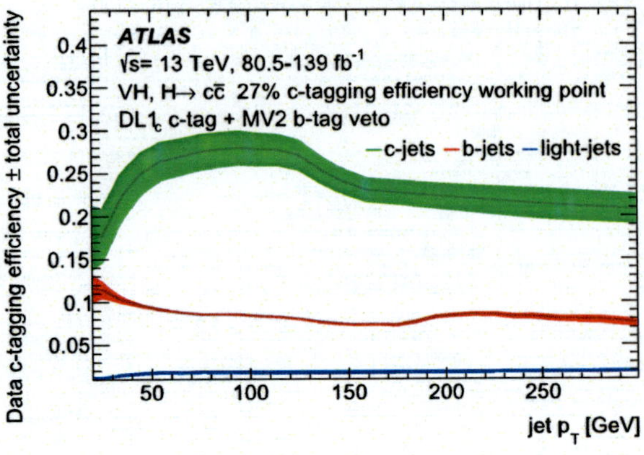

Fig. 4.15 The c-tagging efficiency measured in data for b-, c- and light-jets as a function of jet transverse momentum. The curves correspond to the simulated efficiency multiplied by the smoothed data-to-simulation corrections

the scale factors are of the order of a few percent and are dominated by the precision of the $t\bar{t}$ modelling.

c-jet SF calibrations The c-tagging calibrations for c-jets are derived in semileptonic $t\bar{t}$ events, i.e. in $t\bar{t}$ events where one of the W bosons decays hadronically [81]. Events are selected by requiring 4-jets, *1-lepton* and $E_{\mathrm{T}}^{\mathrm{miss}}$. A kinematic fit is then used to assign the jets to the W boson and top-quark decay [82]. The light- and b-jet scale factors are applied. The c-jet efficiency in data is then extracted by performing a binned χ^2 fit to all categories. The uncertainties on the scale factors are roughly 10% and dominated by the $t\bar{t}$ modelling uncertainties.

τ-jet SF calibrations For τ-jets, the same flavour tagging calibrations are used as for c-jets, with an additional uncertainty of 22% added in each p_T bin.

Light-jet SF calibrations The c-tagging efficiency on light-jets is derived using the negative tag method in a sample of Z+jets events [83]. In the negative tag method, a dedicated tagger is defined, which uses the same quantities as the nominal `DL1` tagger, but where the sign of any signed quantities (e.g. the impact parameter) is inverted. This `DL1Flip` tagger results in a c-tagging efficiency for light jets, which is of the same size as the c-tagging efficiency for c-jets in the nominal tagger, but with significantly improved rejection for heavy-flavour jets. This `DL1Flip` tagger is used for the derivation of the light-jet flavour tagging efficiencies, while the `MV2c10` b-tagger is applied in its nominal configuration. The uncertainties on the light-jet scale factors are roughly 15% for $p_T < 250$ GeV and are dominated mainly by extrapolation uncertainties related to the negative tag approach.

c-tagging efficiencies Finally, the c-tagging efficiencies for b-, c- and light-jets as a function of jet p_T are shown in Fig. 4.15. The efficiencies correspond to the simulated efficiency derived on a $t\bar{t}$ sample multiplied by the smoothed data-to-simulation scale factors.

MC/MC scale factors The MC/MC scale factors are determined by comparing the flavour tagging efficiencies obtained in $t\bar{t}$ samples generated with different gener-

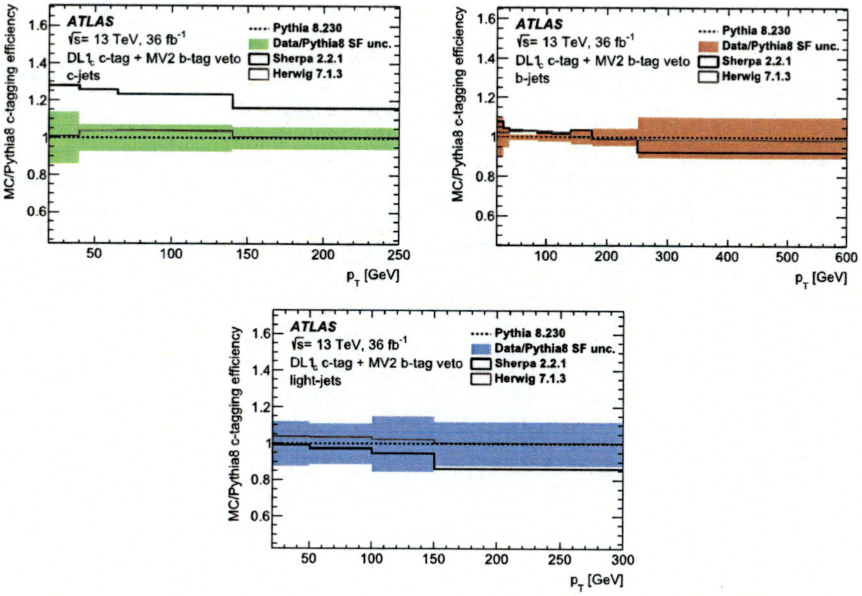

Fig. 4.16 The c-tagging efficiency in different MC samples with respect to the nominal sample used to calibrate the c-tagging performance for c-, b- and light-flavoured jets

ators, with the results shown in Fig. 4.16 as a function of jet momentum for all jet flavours. Along with the MC/MC scale factors, the uncertainty band on the nominal SFs is shown, which are derived in the respective calibration measurements. For c-jets, the MC/MC scale factor for SHERPA 2.2.1 is consistently above one and outside of the error band of the nominal scale factors, which means that the efficiencies derived using PYTHIA 8 are always higher. This is a known feature and can be explained by a sub-optimal baryon fraction parameter in SHERPA 2.2.1, particularly affecting c-jets. For b-jets, the scale factors for SHERPA 2.2.1 are above one at low p_T and below one at high p_T, but the uncertainties largely cover the differences. The scale factors for light-jets in SHERPA 2.2.1 are consistently below one, but within the determined uncertainties.

4.5.2 Truth Tagging

The c-tagger used in this analysis has a modest c-tagging efficiency of 27% and good light and b-jet rejection. Consequently, when applying this tagger to the simulated samples, many events fail the c-tagging requirement and are discarded, which leads to a sizeable statistical uncertainty on the simulated background samples. A parameterised efficiency approach known as *truth tagging* is used to mitigate this. Instead of placing a direct requirement on the c-tagging discriminant, which is referred to as *direct tagging*, events are weighted based on their probability of passing the c-tagging and b-veto requirement.

The event weights are calculated using the finely binned flavour tagging efficiency maps, which are derived as a function of p_T and η. The c-tagging working point includes the b-veto, therefore the efficiency maps are calculated accounting for c-tagger and b-veto. The b-veto on the third jet is applied as a direct cut, and only the two leading jets are considered in the truth-tagging weight computation. Based on the efficiency maps, the truth-tagging weights for the 1 c-tag and 2 c-tag categories are calculated as:

$$\mathrm{TT}_{\mathrm{weight,\ 1\ c-tag}} = \epsilon_1 \times (1 - \epsilon_2) + \epsilon_2 \times (1 - \epsilon_1) \tag{4.5}$$

$$\mathrm{TT}_{\mathrm{weight,\ 2\ c-tag}} = \epsilon_1 \times \epsilon_2, \tag{4.6}$$

where ϵ_1 and ϵ_2 are the c-tagging efficiencies for the first and second jet. Truth-tagging is used for all samples in the analysis, apart from the $VH(H \to c\bar{c})$ signal and $VH(H \to b\bar{b})$ background, which already have sufficient simulation statistics.

ΔR Correction for V+jets

Comparisons of the m_{cc} and ΔR_{cc} distributions for direct and truth tagging show good agreement for $t\bar{t}$ samples, as would be expected as the efficiency maps are derived on $t\bar{t}$ events. For V+jets, discrepancies are observed at low ΔR. For events with a small separation between jets two effects could lead to differences between the truth tagging and direct tagging:

1. At low ΔR, fragmentation in a nearby jet can impact the c-tagged jet. For example, a c-jet could be misidentified as a b-jet if a b-hadron leaked into the c-jet. Generally, additional nearby tracks or vertices can also impact the flavour tagging efficiency.
2. Similarly, at low ΔR between the two jets, the calorimeter clusters of a jet can also be affected by nearby energy depositions from other jets, which then shift the jet axis. This would impact the performance of the `JetFitter` algorithm, which assumes that the jet axis is on the flight path of the heavy hadron.

These effects would be visible in the direct tagging procedure but are not accounted for in the truth tagging calculation. To mitigate this discrepancy at low ΔR, a ΔR dependent correction factor is used. This correction factor is applied per jet and depends on the true flavours of the leading and closest neighbouring jet. The correction factors are derived in a sample of Z+jets events in the *2-lepton* channel by comparing the direct and truth tagged ΔR distributions and fitting a fourth-order polynomial to the ratio of the two. With the additional ΔR correction, the closure between truth tagging and direct tagging was tested once more. A significant improvement of the agreement between direct tagging and truth tagging was found in all cases, with one example shown in Fig. 4.17.

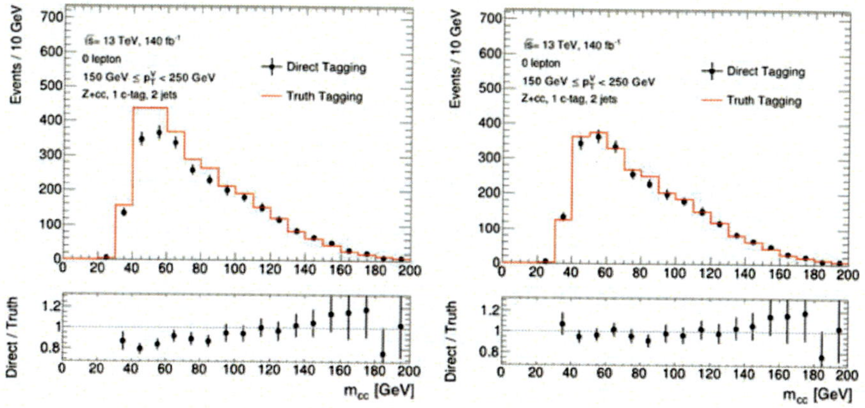

Fig. 4.17 Truth and direct tagging comparison for the invariant mass distribution of a $Z + cc$ sample for the *0-lepton* channel, before (left) and after (right) the ΔR correction

Residual Non-closure Correction

While the ΔR correction significantly improves the agreement between direct tagging and truth-tagging, small residual differences are possible. These differences manifest as an overall normalisation effect. To correct the truth-tagged templates, the normalisations of the truth-tagged templates are scaled to those of the direct tagging templates (*DT correction*). However, in the 2 c-tag categories, the direct tagging templates are often statistically limited, and fluctuations could cause a bias in the corrections. Consequently, the normalisation correction factors are derived in a dedicated likelihood fit as a function of the sample, and the direct and truth-tagging yields.

In summary, *truth-tagging* is applied in the analysis to mitigate the limited statistics of the simulated background samples. The truth tagging weights are calculated based on finely binned flavour tagging efficiency maps, and several corrections are applied to ensure good agreement between direct tagging and truth tagging.

4.6 Experimental Uncertainties

Several areas of experimental uncertainties need to be considered in the $VH(H \to c\bar{c})$ analysis, relating to detector performance, trigger efficiencies, object reconstruction, and flavour tagging. The uncertainties are assessed following the recommendations of the corresponding ATLAS combined performance group. All of the experimental uncertainties are summarised in Table 4.4.

Table 4.4 Experimental uncertainties considered in the $VH(H \to c\bar{c})$ analysis, categorised by their sources

Group	Name	Description
e trigger	EL_EFF_Trigger_Total_1NPCOR_PLUS_UNCOR	Total (statistical and systematic) electron trigger efficiency uncertainty
μ trigger	MUON_EFF_TrigSystUncertainty	Muon trigger efficiency systematic uncertainty
	MUON_EFF_TrigStatUncertainty	Muon trigger efficiency statistical uncertainty
$E_{\mathrm{T}}^{\mathrm{miss}}$ Trigger	MET_Trig_Stat	Statistical error of the $W(\mu, \nu)$+jets used to derive the SFs
	MET_Trig_Top	Account for differences in SF measurements for different $t\bar{t}$ process
	MET_Trig_Z	Account for differences in SF measurements for different Z+jets process
	MET_Trig_SumPt	Account for kinematic dependences on SumPt
e reconstruction	EL_EFF_Reco_TotalCorrUncertainty	Electron reconstruction efficiency total uncertainty
	EL_EFF_ID_TotalCorrUncertainty	Electron ID efficiency total uncertainty
	EL_EFF_Iso_TotalCorrUncertainty	Electron isolation efficiency total uncertainty
	EG_RESOLUTION_ALL	energy resolution total uncertainty
	EG_SCALE_ALL	energy scale total uncertainty
μ reconstruction	MUON_EFF_RECO_STAT	Muon reconstruction efficiency statistical uncertainty (for $p_T > 15\,\mathrm{GeV}$)
	MUON_EFF_RECO_STAT_LOWPT	Muon reconstruction efficiency statistical uncertainty (for $p_T < 15\,\mathrm{GeV}$)
	MUON_EFF_RECO_SYS	Muon reconstruction efficiency systematic uncertainty (for $p_T > 15\,\mathrm{GeV}$)
	MUON_EFF_RECO_SYS_LOWPT	Muon reconstruction efficiency systematic uncertainty (for $p_T < 15\,\mathrm{GeV}$)
	MUON_EFF_ISO_STAT	Muon isolation efficiency statistical uncertainty
	MUON_EFF_ISO_SYS	Muon isolation efficiency systematic uncertainty
	MUON_TTVA_STAT	Muon track-to-vertex association efficiency statistical uncertainty
	MUON_TTVA_SYS	Muon track-to-vertex association efficiency systematic uncertainty
	MUON_SAGITTA_RHO	momentum scale uncertainty to cover charge-dependent local misalignment effects
	MUON_SAGITTA_RESBIAS	momentum scale uncertainty to cover charge-dependent local misalignment effects
	MUONS_SCALE	Muon momentum scale uncertainty
	MUONS_ID	Muon momentum resolution uncertainty (ID component)
	MUONS_MS	Muon momentum resolution uncertainty (MS component)
$E_{\mathrm{T}}^{\mathrm{miss}}$ reconstruction	MET_SoftTrk_ResoPara(Perp)	'soft term'-related longitudinal(transverse) resolution uncertainty
	MET_SoftTrk_Scale	'soft term'-related scale uncertainty
	MET_JetTrk_Scale	$E_{\mathrm{T,Trk}}^{\mathrm{miss}}$ scale uncertainty

(continued)

Table 4.4 (continued)

Group	Name	Description
τ reconstruction	TAUS_TRUEHADTAU_SME_TES_DETECTOR	Tau energy scale uncertainty: single-particle response + threshold
	TAUS_TRUEHADTAU_SME_TES_INSITU	Tau energy scale uncertainty: total from in-situ measurement
	TAUS_TRUEHADTAU_SME_TES_MODEL	Tau energy scale uncertainty: modelling + closure
JET	JET_CR_JET_EffectiveNP_Detector{1-2}	in-situ JES analysis uncertainty
	JET_CR_JET_EffectiveNP_Mixed{1-3}	in-situ JES analysis uncertainty
	JET_CR_JET_EffectiveNP_Modelling{1-4}	in-situ JES analysis uncertainty
	JET_CR_JET_EffectiveNP_Statistical{1-6}	in-situ JES analysis uncertainty
	JET_CR_JET_BJES_Response	b and c-jet energy response uncertainty
	JET_CR_JET_EtaIntercalibration_Modelling	energy scale uncertainties to cover η-intercalibration non-closure
	JET_CR_JET_EtaIntercalibration_NonClosure_highE	energy scale uncertainties to cover η-intercalibration non-closure
	JET_CR_JET_EtaIntercalibration_NonClosure_negEta	energy scale uncertainties to cover η-intercalibration non-closure
	JET_CR_JET_EtaIntercalibration_NonClosure_posEta	energy scale uncertainties to cover η-intercalibration non-closure
	JET_CR_JET_EtaIntercalibration_TotalStat	energy scale uncertainties to cover η-intercalibration non-closure
	JET_CR_JET_Pileup_OffsetMu	pileup uncertainty
	JET_CR_JET_Pileup_OffsetNPV	pileup uncertainty
	JET_CR_JET_Pileup_PtTerm	pileup uncertainty
	JET_CR_JET_Pileup_RhoTopology	pileup uncertainty
	JET_CR_JET_Flavor_Composition	q/g uncertainty
	JET_CR_JET_Flavor_Response	q/g uncertainty
	JET_CR_JET_PunchThrough_MC16	punch-through uncertainty
	JET_CR_JET_SingleParticle_HighPt	high p_T single hadron uncertainties
	JET_CR_JET_JER_DataVsMC	Jet energy resolution total uncertainty
	JET_CR_JET_JER_EffectiveNP_{1-6,7restTerm}	Jet energy resolution total uncertainty
Flavour tagging	FT_EFF_Eigen_Light_{0-3}	Light flavour jet c-tagging efficiency uncertainty
	FT_EFF_Eigen_C_{0-3}	c-jet c-tagging efficiency uncertainty
	FT_EFF_Eigen_B_{0-7}	b-jet c-tagging efficiency uncertainty
	FT_EFF_extrapolation	c-tagging efficiency uncertainty for high-p_T b- and c-jets
	FT_EFF_Eigen_T_{0-3}	τ-jet c-tagging efficiency uncertainty
	FT_EFF_extrapolation_from_charm	τ-jets additional uncertainty extrapolated from c-jets
	TT_dR	truth-tagging ΔR correction uncertainty
	DT_norm	truth-tagging residual non-closure uncertainty

Luminosity and Pile-up

The uncertainty on the measured luminosity in the data taking period between 2015 and 2018 was measured to be 1.7%, following the methodology outlined in [84], from a preliminary calibration of the luminosity scale using x-y beam-separation scans performed during dedicated running periods in each year.

The uncertainty on the pile-up distribution in simulated events is obtained by varying the nominal $< \mu >$ data rescaling factor (1.0/1.09). This factor is used to rescale the μ values (defined in Sect. 3.1.1) in data before calculating the pile-up weights. It is known that the current simulated MC events at a given μ better describe data events at a higher μ, therefore this reweighting is necessary. The associated systematic uncertainty is estimated by varying the rescaling value between 1.0/1.0 and 1.0/1.18 to determine a 1σ uncertainty range.

Triggers: Lepton and $E_{\mathrm{T}}^{\mathrm{miss}}$

In this analysis, lepton and $E_{\mathrm{T}}^{\mathrm{miss}}$ triggers are used to identify events in the different channels. Uncertainties related to the efficiencies of the lepton triggers are included. For electrons, one uncertainty is considered, covering both statistical and systematic uncertainties. For muons, the statistical and systematic uncertainties are considered separately.

For the $E_{\mathrm{T}}^{\mathrm{miss}}$ trigger, scale factors to account for the trigger efficiency turn-on are derived from W+jets events and have been studied in detail in the $VH(H \to b\bar{b})$ analysis [85]. Several sources of uncertainties are considered for the scale factors. First, the statistical uncertainty on the dataset used to derive the scale factors is considered. Second, a systematic uncertainty is derived using a different sample ($t\bar{t}$ or Z+jets) to compute the scale factors. A third uncertainty accounts for the dependence of the $E_{\mathrm{T}}^{\mathrm{miss}}$ trigger efficiency on the scalar sum of all final state jets within the event.

Lepton and $E_{\mathrm{T}}^{\mathrm{miss}}$ Reconstruction

Uncertainties related to the reconstruction of leptons and $E_{\mathrm{T}}^{\mathrm{miss}}$ are included in the $VH(H \to c\bar{c})$ analysis and propagated from the calibration analyses of these objects.

The electron reconstruction and identification efficiency and the associated systematic uncertainties are derived by comparing data and simulated samples in $Z \to e^+e^-$, $W \to e\nu$ and $J/\psi \to e^+e^-$ events and are provided in bins of p_T and $|\eta|$. Uncertainties related to the reconstruction efficiency itself, the identification efficiency scale factors and the isolation efficiency scale factors are provided [86]. In the $VH(H \to c\bar{c})$ analysis regime, these uncertainties are of the order of a few percent. In the electron energy calibration, over 60 different systematic uncertainties are considered. Due to the small impact on this analysis, only two uncertainties are included, which group this large number of uncertainties together based on their impact on the electron energy scale and resolution.

The muon reconstruction scale factors are calibrated in samples of $Z \to \mu^+\mu^-$ and $J/\psi \to \mu^+\mu^-$ events in the 2015–2018 dataset. Uncertainties related to the reconstruction and isolation efficiency, and track-to-vertex association are considered. For low p_T muons the J/ψ measurement is more critical, and for high p_T muons the Z measurement is more important, and separate uncertainties are provided for both regimes. Similarly, the muon energy scale and resolution have been calibrated and the related uncertainties are also included in the analysis.

Uncertainties related to the $E_{\mathrm{T}}^{\mathrm{miss}}$ reconstruction scale and resolution are included.

Jet Energy Scale and Resolution

In the jet calibration analysis, a full set of around 50 systematic uncertainties is derived. A globally reduced set of uncertainties is also provided. It covers uncertainties related to *in-situ* analyses, η intercalibration, high-p_{T} jets, pile-up, flavour composition, flavour response, b-jets, and punch-through jets. This reduced set of uncertainties is used in the $VH(H \rightarrow c\bar{c})$ analysis. It is derived by performing a principal component analysis and combining the resulting variations in such a way as to preserve correlations in certain regions of jet kinematics. Uncertainties related to the jet energy scale (JES) and jet energy resolution (JER) are included.

Flavour Tagging

The calibration of the flavour tagging scale factors is described in Sect. 4.5.1 and there are approximately 20 different systematic uncertainties related to the c-tagging for each jet flavour. A principal component analysis is performed to reduce the number of systematic uncertainties. For each jet flavour, an uncertainty is included for every jet momentum bin in the flavour tagging scale factor calibration. By default, the systematic uncertainties for the τ-jets are grouped with the ones of c-jets, and one separate extrapolation uncertainty is added. The *0-lepton* channel has a sizeable τ-jet contribution; thus, the uncertainties related to the flavour tagging were chosen to be fully decorrelated between the c- and τ-jets in this analysis.

Truth Tagging

The systematic uncertainties related to the truth tagging efficiencies are expected to be covered in the systematics of the calibration scale factors, as the nominal flavour tagging efficiency maps are used. In addition, two systematic uncertainties are added to account for residual non-closure between truth and direct tagging. This first systematic uncertainty accounts for the ΔR correction applied to truth tagging. It is taken to be the total size of the correction, considering both its effect on the shape and normalisation of the m_{cc} distribution. The systematic uncertainty is derived separately for each jet and close-by jet flavour. The second systematic uncertainty relates to the residual non-closure correction and is implemented as the full size of the normalisation correction.

4.7 Signal and Background Modelling Uncertainties

This section describes the uncertainties derived for the various signal and background processes present in the $VH(H \rightarrow c\bar{c})$ analysis.

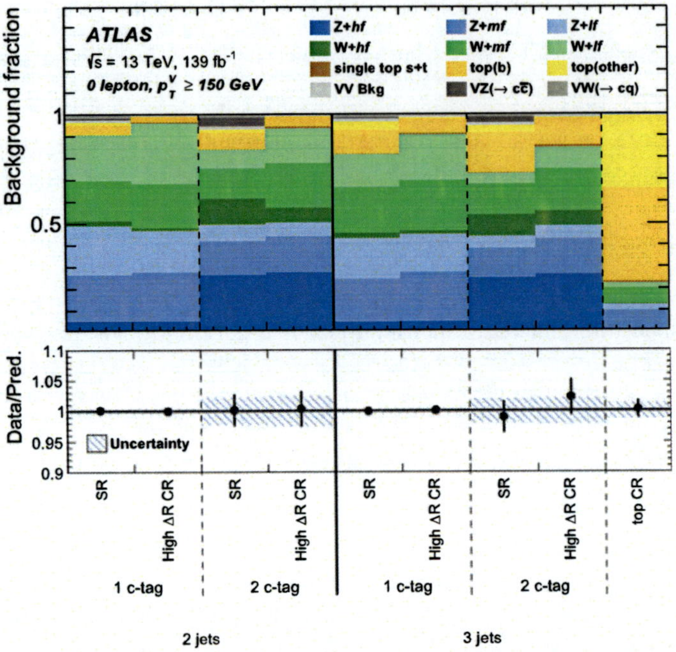

Fig. 4.18 The background composition in all signal and control regions of the *0-lepton* channel

4.7.1 Background Composition in the Different Channels

Before studying the background modelling uncertainties, it is necessary to understand the background composition in the different channels. The background compositions in all three channels, in all signal and control regions are shown in Figs. 4.18, 4.19 and 4.20.

***0-lepton* channel** The *0-lepton* channel is affected by a very diverse background composition in its signal regions. The most considerable background comes from the V+jets processes, with equal Z+jets and W+jets contributions. Z+jets events enter the *0-lepton* channel when the Z boson decays into two neutrinos. The W+jets process passes the selection criteria in events with high $E_{\mathrm{T}}^{\mathrm{miss}}$, where the lepton is not correctly identified, or when the W boson decays into a hadronically decaying τ-lepton. The second-largest background in the *0-lepton* channel is the top quark background, which is more sizeable in the 3-jet signal region. The top quark background is constrained in its dedicated control region, which is very pure in $t\bar{t}$ and single top events, with only a negligible contribution of V+jets. Finally, the diboson processes provide a small contribution to the total background to the $VH(H \to c\bar{c})$ signal.

***1-lepton* channel** The *1-lepton* channel mainly contains W+jets and top quark backgrounds. The relative composition of the W+jets events is similar to the *0-lepton*

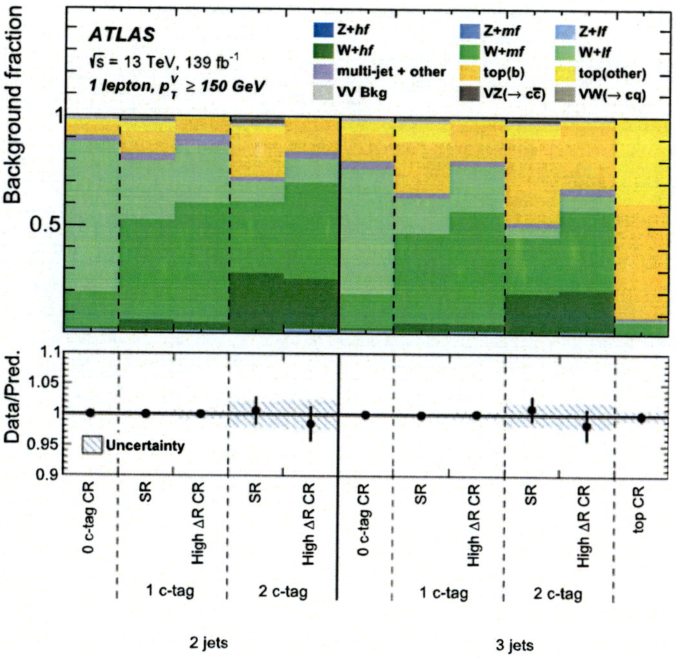

Fig. 4.19 The background composition in all signal and control regions of the *1-lepton* channel

channel. In the *1-lepton* channel, 0 c-tag control regions are included, which primarily contain W+light jet events. Compared to the *0-lepton* channel, the top background contribution in the signal regions is significantly larger in the *1-lepton* channel, constituting almost half of the event in the 2 c-tag 3-jet signal region. Once again, the top control region is very pure in top quark background events. Finally, QCD multi-jet and diboson events constitute a minor background to the $VH(H \to c\bar{c})$ signal.

***2-lepton* channel** The background composition in the *2-lepton* channel is significantly more straightforward. Here, the signal regions contain mainly Z+jets events, and a 0 c-tag control region is included for constraining the Z+light jet background. In the low p_T^V categories, there is a sizeable contribution of $t\bar{t}$ events, which is constrained in dedicated, very pure, $e\mu$ control regions. Once again, the diboson events are a minor background in the signal regions of the *2-lepton* channel.

There is a small contribution of $VH(H \to b\bar{b})$ events in all channels, which is not large enough to be visible in Figs. 4.18, 4.19 and 4.20.

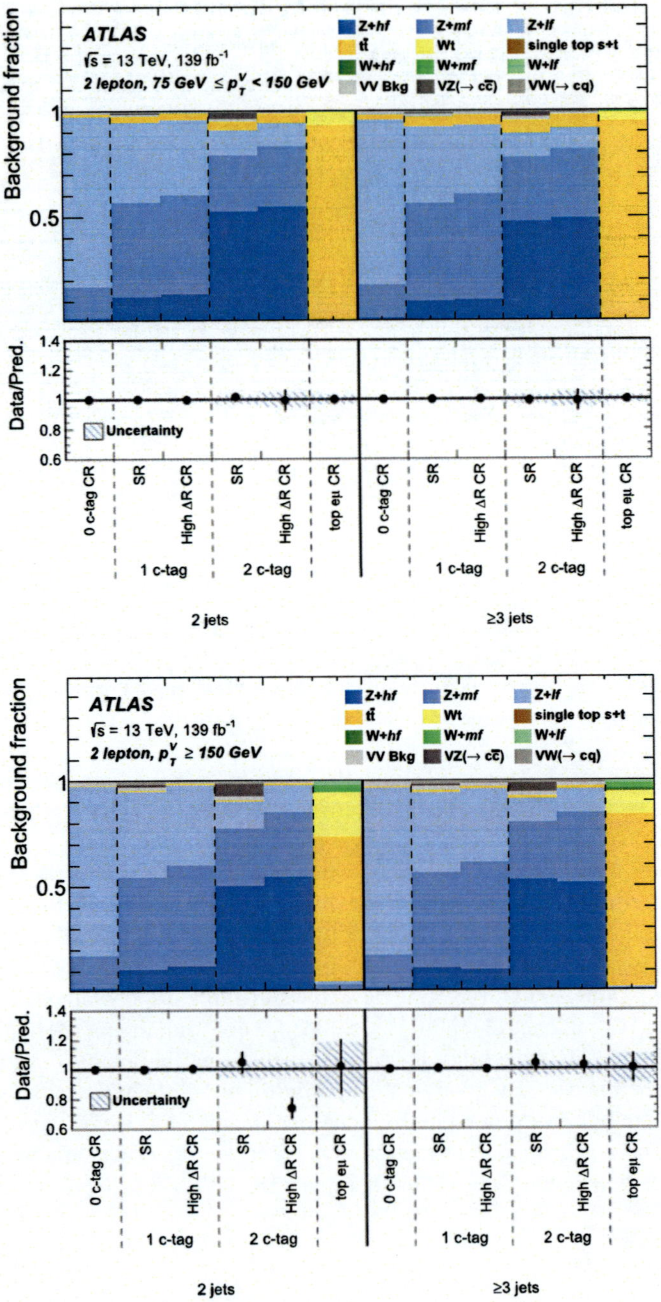

Fig. 4.20 The background composition in all signal and control regions of the *2-lepton* channel for events with $75 < p_T^V < 150$ GeV (top) or $p_T^V > 150$ GeV (bottom)

Table 4.5 Summary of samples in the analysis, their nominal generators and the alternative generators and scale variations used to assess signal and background modelling uncertainties

Sample	Nominal generator	Alternative generators	Scale variations
$VH(H \to c\bar{c})$	POWHEG + PYTHIA 8	POWHEG + HERWIG 7	μ_R, μ_F
$VH(H \to b\bar{b})$	POWHEG + PYTHIA 8	POWHEG + HERWIG 7	μ_R, μ_F
Diboson	SHERPA 2.2.1	POWHEG +PYTHIA 8	μ_R, μ_F
$t\bar{t}$ and single top	POWHEG +PYTHIA 8	POWHEG +HERWIG 7, MADGRAPH 5_aMC@NLO+PYTHIA 8.2	Additional initial/final state radiation
V+jets	SHERPA 2.2.1	MADGRAPH 5	μ_R, μ_F

4.7.2 *Derivation of Signal and Background Modelling Uncertainties*

The signal and background processes in the $VH(H \to c\bar{c})$ analysis are mostly modelled using simulated samples. To assess the uncertainties on the generated samples, the *nominal samples* are compared to *alternative samples* differing in matrix element generation, perturbation theory corrections, or parton shower generation. Additionally, variations of the renormalisation and factorisation scales are considered. A summary of the sources of uncertainties for the different samples is summarised in Table 4.5. When comparing these nominal and alternative generators, different kinds of uncertainties can be derived. An overview of the derived uncertainties can be found in Table 4.6, and details about the treatment of individual processes can be found in Sects. 4.7.3–4.7.8.

Normalisations The overall normalisations of background processes (across several or all analysis categories) can either be measured in data, when the normalisations are entirely free to float in the fit, or determined from the simulated samples, with an associated normalisation uncertainty determined from generator comparisons.

Acceptance ratios The predicted yields in different analysis regions (for example, p_T^V, jet multiplicity bins, or signal and control regions) can vary depending on which MC generator is used. Acceptance uncertainties are introduced to account for these different predictions. These uncertainties are derived as double ratios between the yields predicted by the nominal and alternative generators in regions 1 and 2:

$$\text{Acceptance ratio} = \sqrt{\sum_i \left(\frac{(\frac{n_1}{n_2})_i}{(\frac{n_1}{n_2})_{\text{nominal}}} - 1 \right)^2}, \tag{4.7}$$

where $(\frac{n_1}{n_2})_{\text{nominal}}$ represents the yield ratio using the nominal MC prediction, and $(\frac{n_1}{n_2})_i$ the yield ratio using the alternative sample. The acceptance ratios are always derived for the region with the highest purity of a particular process, designated as

Table 4.6 Summary of the background modelling systematic uncertainties considered. The values given refer to the size of the uncertainty affecting the yield of each background. Uncertainties in the shapes of the m_{cc} distributions are not shown below, but are taken into account for all backgrounds

$VH(\rightarrow b\bar{b})$	
$WH(\rightarrow b\bar{b})$ normalisation	27%
$ZH(\rightarrow b\bar{b})$ normalisation	25%
Diboson	
$WW/ZZ/WZ$ acceptance	10%/5%/12%
p_T^V acceptance	4%
N_{jet} acceptance	7–11%
Z+jets	
$Z+hf$ normalisation	Floating
$Z+mf$ normalisation	Floating
$Z+lf$ normalisation	Floating
$Z+bb$ to $Z+cc$ ratio	20%
$Z+bl$ to $Z+cl$ ratio	18%
$Z+bc$ to $Z+cl$ ratio	6%
p_T^V acceptance	1–8%
N_{jet} acceptance	10–37%
High-ΔR CR to SR	12–37%
0- to *2-lepton* ratio	4–5%
W+jets	
$W+hf$ normalisation	Floating
$W+mf$ normalisation	Floating
$W+lf$ normalisation	Floating
$W+bb$ to $W+cc$ ratio	4–10%
$W+bl$ to $W+cl$ ratio	31–32%
$W+bc$ to $W+cl$ ratio	31–33%
$W \rightarrow \tau\nu(+c)$ to $W+cl$ ratio	11%
$W \rightarrow \tau\nu(+b)$ to $W+cl$ ratio	27%
$W \rightarrow \tau\nu(+l)$ to $W+l$ ratio	8%
N_{jet} acceptance	8–14%
High-ΔR CR to SR	15–29%
$W \rightarrow \tau\nu$ SR to high-ΔR CR ratio	5–18%
0- to *1-lepton* ratio	1–6%
Top quark (*0-* and *1-lepton*)	
Top(b) normalisation	Floating
Top(other) normalisation	Floating
N_{jet} acceptance	7–9%
0- to *1-lepton* ratio	4%
SR/top CR acceptance ($t\bar{t}$)	9%
SR/top CR acceptance (Wt)	16%
Wt / $t\bar{t}$ ratio	10%
Top quark (*2-lepton*)	
Normalisation	Floating
Multi-jet (*1-lepton*)	
Normalisation	20–100%

region 2. The uncertainty is then applied to the less pure region, denoted as region 1. For example, an acceptance uncertainty can be derived for extrapolating from a control region to a signal region: The double ratio of SR/CR yields is calculated for nominal and alternative generators. The uncertainty is then applied to the signal region. Usually, several different sources of uncertainty are considered, and the uncertainties from each source are added in quadrature, as illustrated in Eq. 4.7.

m_{cc} shape uncertainties Uncertainties on the predictions of the m_{cc} shape are also calculated by comparing nominal and alternative generators. Here, the m_{cc} distributions are usually normalised, and the ratio of nominal and alternative generators is parametrised in a functional form to smooth out statistical fluctuations. This parametrisation is then used as an uncertainty in the analysis.

Detailed studies have been carried out to derive the signal and background modelling uncertainties. Generally, the uncertainties are first calculated with categories split as finely as possible, in the number of c-tags, jet multiplicity, and p_T^V. Then, regions are combined to provide better statistics if the initially derived uncertainties are compatible. In general, the same uncertainties are used in different channels and regions, as long as they are within the statistical uncertainties of the simulated samples.

4.7.3 V+Jets Background

V+jets processes constitute the most significant background in the $VH(H \rightarrow c\bar{c})$ analysis. Z+jets and W+jets events are dominant in the *2-lepton* and *1-lepton* channels respectively, while the *0-lepton* channel includes a combination of both. The W+jets and Z+jets processes are treated similarly in deriving the systematic uncertainties. The Z+jets and W+jets backgrounds are initially divided into components by di-jet flavour, e.g. $Z + bb$, $Z + bc$, $Z + bl$, $Z + cc$, $Z + cl$ and $Z + l$. For Z+jets, the flavour composition of the 1 and 2 c-tag regions in the *2-lepton* channel is shown in Fig. 4.21 and a similar flavour composition can also be found in the *0-lepton* channel. The 1 c-tag categories are dominated by $Z + cl$ and $Z + l$, and the 2 c-tag categories mainly consist of $Z + cc$ events.

The flavour composition of the W+jets background is very similar to the that of Z+jets. In addition to the di-jet flavours listed above, events with hadronically decaying taus are also considered in the *0-lepton* channel: $W(\tau\nu) + b$, $W(\tau\nu) + c$ and $W(\tau\nu) + l$. The composition is consistent between the channels and shows similar relative contributions in the 1 c-tag and 2 c-tag categories. However, in the *0-lepton* channel, there are up to 10% of W+jets events with τ-jets.

Normalisations Based on the di-jet flavour, the Z+jets and W+jets backgrounds are grouped into different categories to combine similar physics processes:

- **W+hf and Z+hf**: Heavy flavour (bb, cc) jets in association with a vector boson are produced through gluon splitting, and the fusion of b/c-quarks. The modelling of these processes is problematic due to significant differences in the simulation

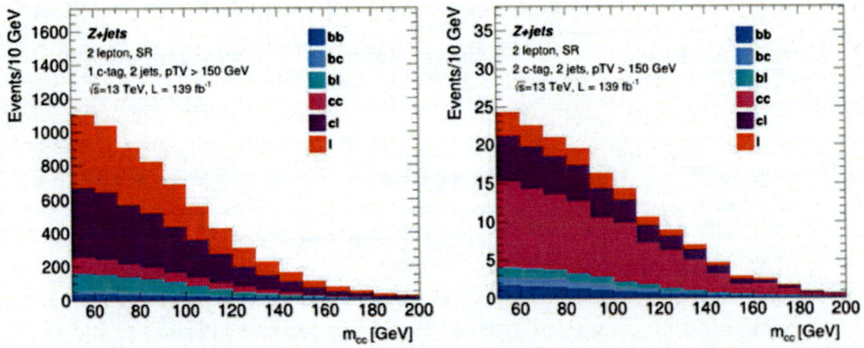

Fig. 4.21 Flavour composition of the Z+jets background in the *2-lepton* channel in the 1 c-tag (left) and 2 c-tag (right) categories

of b- and c-based samples. Consequently, these backgrounds are merged into one as $W + hf\{bb, cc\}$ and $Z + hf\{bb, cc\}$.

- Z+mf and W+mf: Mixed flavour events with a combination of heavy- and light-flavour quarks are usually produced similarly, so these components are grouped into a single $Z + mf\{bc, bl, cl\}$ and $W + mf\{bc, bl, cl\}$ component. In the *0-lepton* channel, this component also includes the $W(\tau\nu) + b$ and $W(\tau\nu) + c$ contributions.
- $Z + l$ and $W + l$: Events with light jets produced in association with a vector boson are treated separately in a $Z + l$ and $W + l$ component. In the *0-lepton* channel this also includes $W(\tau\nu) + l$ events.

All Z+jets and W+jets background normalisations are left free to float in the fit. The $Z + l$ and $W + l$ backgrounds are constrained using a dedicated 0 c-tag control region in the *2-lepton* and *1-lepton* channel, respectively. There are six single-bin control regions introduced in the fit, split by the p_T^V and jet multiplicity categories. Due to the high available statistics of the 0 c-tag region, each control region has one associated floating normalisation, i.e. decorrelated for each p_T^V and jet multiplicity bin. The V+hf and V+mf components are constrained from the high ΔR control region and the signal region itself. The Z+hf and W+hf contributions are mainly composed of $Z + cc$ events; therefore, they will be best constrained by the 2 c-tag categories. In contrast, the Z+mf and W+mf backgrounds are primarily composed of $Z + cl$; thus, they will be constrained by the 1 c-tag categories. The Z+hf and Z+mf floating normalisations are decorrelated between different p_T^V categories to allow some additional freedom for the low p_T^V category in *2-lepton* with respect to the $p_T^V > 150$ GeV categories in *0-lepton* and *2-lepton*.

Acceptance Uncertainties Acceptance ratios are estimated for both the SRs and ΔR CRs and computed between the different categories: jet multiplicity, flavour composition and channels. As the floating normalisations are decorrelated in p_T^V, no acceptance ratios are applied between p_T^V categories. Additionally, in the *0-lepton*

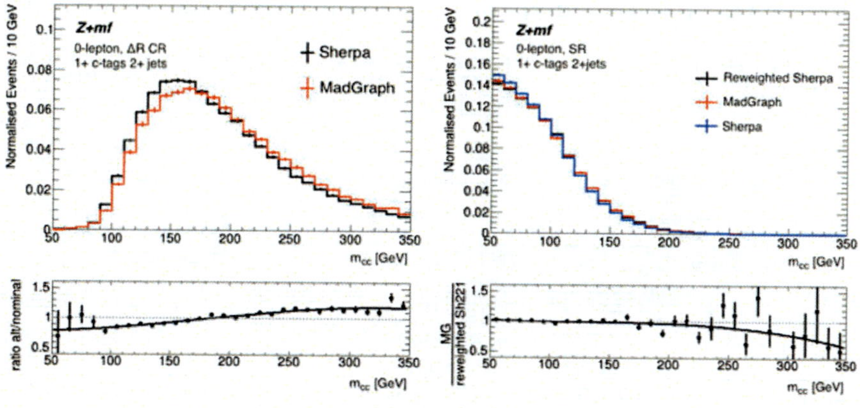

Fig. 4.22 Examples of the m_{cc} shape systematics derived by comparing the m_{cc} shapes in samples generated using SHERPA and MADGRAPH for Z+mf events. *Systematic 1* (left) is derived in the ΔR CR and *Systematic 2* (right) is derived in the SR

channel, uncertainties between the SR and ΔR CR are calculated for the $W(\tau\nu)$+jets components.

Shape Uncertainties The m_{cc} shape uncertainties are used to parametrise the uncertainty on the m_{cc} distributions in the signal and control regions and take into account the acceptance uncertainty between SR and ΔR CR. They are derived as two different sets of shape systematics. The first one, labelled as *Systematic 1*, is computed by parametrising the ratio of the m_{cc} shapes in samples generated using SHERPA and MADGRAPH in the high ΔR CR only. It is applied as a shape and normalisation uncertainty to the SR and ΔR CR. This allows to extrapolate the m_{cc} shape from the ΔR CR to the SR and takes care of the acceptance effect between the two regions. The second systematic uncertainty (*Systematic 2*) is added to introduce additional freedom on the m_{cc} shape in the SR. This uncertainty is derived by reweighting the nominal m_{cc} distributions by *Systematic 1*. Then the ratio between the reweighted m_{cc} distribution and the alternative MADGRAPH m_{cc} distribution in the signal region is calculated, parametrised and used as the second systematic uncertainty. The uncertainties are derived separately for W+jets and Z+jets and split into the V+hf, V+mf and $V + l$ components. Additionally, for $W + l$ and $Z + l$, distinct shapes are used for the 2-jet and 3-jet categories, as the calculated m_{cc} ratios show different trends. An example of the m_{cc} shape uncertainties derived for Z+jets is shown in Fig. 4.22.

Additional m_{cc} shape uncertainties are included to account for variations in the μ_R and μ_F scale in the nominal samples. These variations are used as a direct input in the analysis, without additional parametrisation using a polynomial.

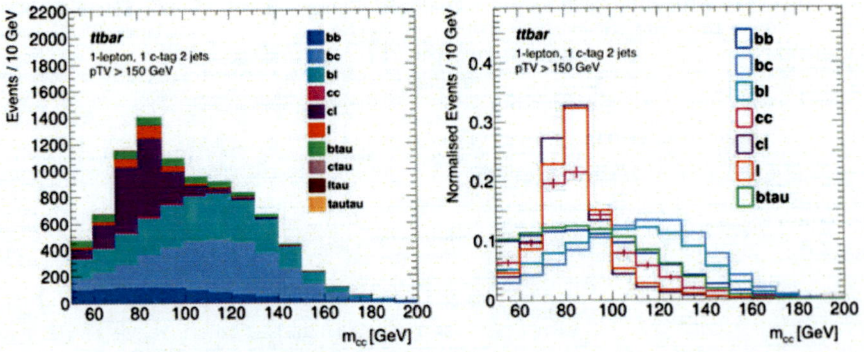

Fig. 4.23 Flavour composition (left) and normalised distributions (right) of the $t\bar{t}$ m_{cc} distributions in the *1-lepton* channel, split by di-jet flavour

4.7.4 Top Quark Background

The backgrounds originating from decays of top quarks, i.e. $t\bar{t}$ and single top, are considered together in the derivation of the modelling systematics. Single top processes are a minor background in the *0-lepton* and *1-lepton* channels, with the main component coming from the production in the Wt-channel. In contrast, the t-channel and s-channel are subdominant, contributing less than 1% to the total background. Consequently, for all three production channels, a cross-section uncertainty is included in the fit, but only for the Wt component additional modelling systematics are considered.

Acceptance ratios and shape uncertainties are derived for the top quark backgrounds in the *0-lepton* and *1-lepton* channels. In the *2-lepton* channel, the contribution of top quark backgrounds is small and directly constrained from a dedicated $e\mu$ control, so no additional acceptance or shape uncertainties are included.

Normalisations The different di-jet flavour components in the top quark background are grouped into two categories, based on their m_{cc} shapes and physics processes, for both $t\bar{t}$ and single top Wt. Figure 4.23 shows the total and normalised m_{cc} distributions for the different di-jet flavour components. Based on this, the following categories are defined:

- **Top(other)**: $(cl + l + cc)$—Both jets are produced in the decay of the W-boson, accordingly, m_{cc} is peaked around 80 GeV.
- **Top(b)**: $(bc + bl + bb + b\tau)$—Here, one of the b-jets of the $t\bar{t}$ decay is reconstructed together with a c/l-jet from the W-decay. This component is peaked around 120 GeV.

Based on this categorisation, two floating normalisations are defined in the *0-lepton* and *1-lepton* channels: the Top(b) and Top(other) normalisations are floated separately, where the $t\bar{t}$ and Wt contributions are merged into one component. The motivation for this treatment is that the Top(b) component is peaked at 120 GeV, very close to the Higgs mass, and the Top(other) is peaked at 80 GeV, i.e. directly under the diboson signals. Floating the different top quark background components separately allows decoupling the two regions in m_{cc}.

In the *2-lepton* channel, the $t\bar{t}$ background is constrained using the top eμ control region, introduced into the fit as a single-bin region for each p_T^V and jet multiplicity category. Each control region has its separate floating normalisation due to the high available statistics in the control region, leading to a total of four normalisations, decorrelated by p_T^V and jet multiplicity category.

Acceptance Uncertainties Acceptance uncertainties are calculated between the top CR and SR, between jet multiplicity regions and between channels. Finally, as Wt and $t\bar{t}$ are grouped into one component, an uncertainty is considered on the relative composition, however only for the Top(b) component, as the Top(other) components are not distinguishable between Wt and $t\bar{t}$, as the latter originates from the W boson decay, which is the same for Wt and $t\bar{t}$.

Shape Uncertainties m_{cc} shape uncertainties are derived for all alternative generators, separately for the Top(b) and Top(other) components and individually for Wt and $t\bar{t}$. The uncertainties are later correlated between Wt and $t\bar{t}$ in the fit. The high ΔR control region is also considered for the Top(b) component when deriving the shape uncertainties. For Top(b), the m_{cc} shapes are derived inclusively in SR and high ΔR CR and applied as a shape and normalisation uncertainty to account for the acceptance effect between the SR and ΔR CR. The remaining $t\bar{t}$ and Wt components are negligible in the ΔR CR, and therefore the m_{cc} shape uncertainties are derived in the SR and top CR only.

4.7.5 Diboson

Diboson processes constitute a small background in the analysis. However, some of the diboson events are later used as a signal to cross-check the analysis strategy. Four different diboson components can enter the analysis:

- ZZ, where one of the Z-bosons decays leptonically and one hadronically
- $W_{lep}Z_{had}$, where the W-boson decays leptonically, and the Z-boson decays hadronically
- $Z_{lep}W_{had}$, where the Z-boson decays leptonically, and the W-boson decays hadronically
- WW, where one of the W-bosons decays leptonically and one hadronically.

Examples of the flavour composition of the different diboson processes are shown in Fig. 4.24. The ZZ component mostly has di-jet flavour cc, and a similar conclusion can also be drawn for the $W_{lep}Z_{had}$ component. The WW component mainly has di-jet flavour cl from the decay of the W boson, and the same is the case for $Z_{lep}W_{had}$.

The different diboson components are then further categorised into signal and background components based on their di-jet flavour:

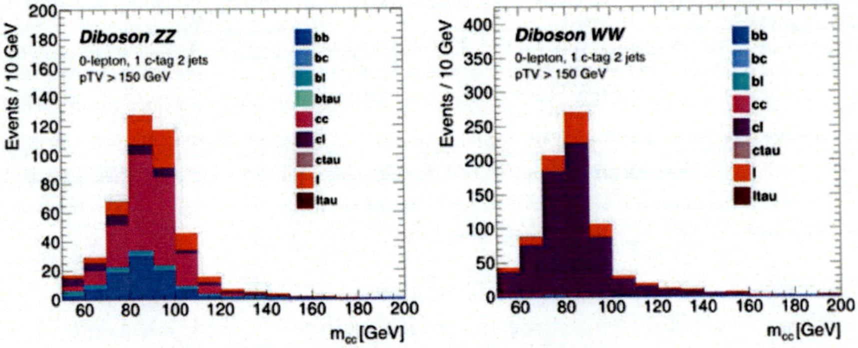

Fig. 4.24 Flavour composition of the m_{cc} distributions of the different diboson components in the 1 c-tag, 2-jet, p_T^V >150 GeV category of the *0-lepton* channel

- $VZ(Z \rightarrow c\bar{c})$: $ZZ(cc)$, $W_{lep}Z(cc)$, $ggZZ(cc)$
- $VW(W \rightarrow cq)$: $WW(cl)$, $Z_{lep}W(cl)$, $ggWW(cl)$
- VV background: all remaining diboson components.

The $VZ(Z \rightarrow c\bar{c})$ and $VW(W \rightarrow cq)$ processes are treated as signals in the analysis. Due to their decays into c-quarks, their measurements provide an excellent cross-check of the $VH(H \rightarrow c\bar{c})$ analysis, particularly the c-tagging. The $VZ(Z \rightarrow c\bar{c})$ and the $VW(W \rightarrow cq)$ processes are relevant for the 2 c-tag and 1 c-tag categories, respectively.

Normalisations A flavour inclusive normalisation uncertainty is derived for the three diboson components WW, WZ and ZZ.

Acceptance Uncertainties In addition to the normalisation uncertainties, acceptance uncertainties are introduced between the different analysis categories, namely, between the low p_T^V and p_T^V > 150 GeV region (in the *2-lepton* channel only) and the 3(+)-jet and 2-jet categories. The three diboson components WW, WZ and ZZ **are considered separately, but the acceptance ratios are derived inclusively in flavour.**

Shape Uncertainties m_{cc} shape uncertainties are derived for the different diboson signal and background components in the signal regions. Polynomials are used to parametrise the ratio of the m_{cc} distributions between the nominal and alternative samples.

4.7.6 $VH(H \rightarrow c\bar{c})$ Signal

A full set of normalisation, acceptance and shape uncertainties is derived for the $VH(H \rightarrow c\bar{c})$ signal.

Normalisations Two uncertainties on the signal normalisation are implemented. Firstly, theory uncertainties on the $H \rightarrow c\bar{c}$ branching ratio, and the ZH and WH production cross-sections are considered. Secondly, an uncertainty on the phase space acceptance is derived from comparing the nominal to the alternative samples. This uncertainty is derived split by the production process.

Acceptance Uncertainties In addition to the normalisation uncertainties, acceptance uncertainties are introduced between the different analysis categories. The three production processes $qqZH(H \rightarrow c\bar{c})$, $qqWH(H \rightarrow c\bar{c})$ and $ggZH(H \rightarrow c\bar{c})$ are considered separately.

Shape Uncertainties m_{cc} shape uncertainties are again derived for the different production processes $qqZH(H \rightarrow c\bar{c})$, $qqWH(H \rightarrow c\bar{c})$ and $ggZH(H \rightarrow c\bar{c})$, which are correlated to a single $VH(H \rightarrow c\bar{c})$ shape uncertainty in the fit.

4.7.7 $VH(H \rightarrow b\bar{b})$ Background

For the $VH(H \rightarrow b\bar{b})$ background, a normalisation uncertainty is applied based on the uncertainty of the signal strength measured in the $VH(H \rightarrow b\bar{b})$ analysis [77]. A separate uncertainty is applied for the $WH(H \rightarrow b\bar{b})$ and $ZH(H \rightarrow b\bar{b})$ components. In addition to this, p_T^V and jet multiplicity acceptance uncertainties are applied, with the same values as for the $VH(H \rightarrow c\bar{c})$ signal. Finally, m_{cc} shape uncertainties are derived in the same way as for the $VH(H \rightarrow c\bar{c})$ signal.

4.7.8 QCD Multi-jet

The QCD multi-jet background is estimated in a dedicated control region, which is used to derive a data-driven template for the multi-jet m_{cc} shape and normalisation. Uncertainties on the multi-jet background are derived separately in the electron and muon channels. Uncertainties on the m_{cc} shape are derived by considering different sources of uncertainty: varying the triggers in the electron channel, the selection of the multi-jet control regions and the W+jets and $t\bar{t}$ normalisations in the control region. Uncertainties on the normalisations are also derived using these same sources of uncertainty. Additional uncertainties on the multi-jet normalisation are determined by using a different distribution ($\Delta\phi(l, MET)$) to perform the template fit, including the $E_{\mathrm{T}}^{\mathrm{miss}} < 30$ GeV region, and applying additional cuts on the multi-jet control region (m_T^W <120 GeV).

4.8 Statistical Analysis

To measure the $VH(H \rightarrow c\bar{c})$ and diboson signals, a *binned likelihood fit* is performed to the invariant mass m_{cc} in all signal and control regions, together in all three channels. The parameters of interest in the fit are the signals strengths of the $VH(\rightarrow c\bar{c})$, $VZ(\rightarrow c\bar{c})$ and $VW(\rightarrow cq)$ processes. The technical implementation of the fit is done using the `WorkSpaceMaker` framework building upon `HistFactory` [87] together with `RooFit` [88] and `RooStats` [89].

4.8.1 Definition of the Likelihood Function

The fit is performed by maximising the *likelihood function* as a function of both the signal strength and the systematic uncertainties, which are implemented as nuisance parameters. The definition of the likelihood function is best explained in several steps. The first part of the likelihood function consists of Poissonian probability terms, based on the expected and observed event yields in all of the bins:

$$\mathscr{L}(\mu) = \prod_{i \in \text{bins}} \text{Pois}(N_i | \mu s_i + b_i) = \prod_{i \in \text{bins}} \frac{(\mu s_i + b_i)^{N_i}}{N_i!} e^{-(\mu s_i + b_i)}, \tag{4.8}$$

where N_i is the number of measured data events in bin i, and s_i, and b_i are expected signal and background counts predicted by simulation. The parameter μ is the signal strength, which is defined as the measured $\sigma \times$ BR for the signal process divided by the Standard Model expectation. In the case of multiple parameters of interest, the signal strengths can be written as a vector $\vec{\mu}$. In this analysis, three signal strengths for the processes $VH(H \to c\bar{c})$, $VZ(Z \to c\bar{c})$ and $VW(W \to cq)$ are measured simultaneously.

Next, the systematic uncertainties are introduced into the fit in the form of nuisance parameters (NPs) $\vec{\theta}$, which can modify the expected signal and background yields: $\{s_i, b_i\} \to \{s_i(\vec{\theta}), b_i(\vec{\theta})\}$. They are implemented as a set of Gaussian penalty terms, which have a mean value of zero and variance of 1:

$$\mathscr{L}_{\text{Aux}} = \prod_{\theta \in \vec{\theta}} \frac{1}{\sqrt{2\pi}} e^{-\theta^2/2}. \tag{4.9}$$

The effect of the nuisance parameters is determined in dedicated auxiliary measurements, as described in Sects. 4.6 and 4.7. For example, a prior of 5% can be implemented for a particular background normalisation uncertainty. If the NP is moved upwards by 1 standard deviation, the yields for this particular background will be increased by 5%. The deviation of the best fit value of a nuisance parameter from zero is the *pull* and a decrease in the variance after performing the fit is the *constraint*. The constraint indicates the certainty with which the value of the NP is known. It is also possible to have NPs without a predetermined prior—these are called *free-floating* parameters and are usually the normalisations of the major backgrounds, which are determined purely from data. The pre-fit values of these floating parameters are set to 1.

Finally, uncertainties related to the limited available statistics of the simulation samples are introduced as γ-parameters. One parameter is introduced per bin, and can modify the expected background yield in that particular bin as $b_i(\vec{\theta}) \rightarrow \gamma_i b_i(\vec{\theta})$. In the likelihood description, these terms are introduced as

$$\mathcal{L}_{\text{BkgStat}}(\vec{\gamma}) = \prod_{i \in \text{bins}} \text{Gauss}(\beta_i | \gamma_i \beta_i, \sqrt{\gamma_i \beta_i}), \tag{4.10}$$

with

$$\beta_i = \frac{1}{\sigma_{\text{rel}}^2}, \tag{4.11}$$

where σ_{rel} is the relative statistical uncertainty on the expected total background yield. The full likelihood can be (schematically) written as

$$\mathcal{L}(\vec{\mu}, \vec{\theta}, \vec{\gamma}) = \prod_{i \in \text{bins}} \text{Pois}(N_i | \vec{\mu} s_i(\vec{\theta}) + \gamma_i b_i(\vec{\theta})) \times \prod_{\theta \in \vec{\theta}} \frac{1}{\sqrt{2\pi}} e^{-\theta^2/2} \times \prod_{i \in \text{bins}} \text{Gauss}(\beta_i | \gamma_i \beta_i, \sqrt{\gamma_i \beta_i}). \tag{4.12}$$

Once constructed, the likelihood is maximised to determine the parameters of interest and their uncertainties. To test a certain hypothesis for the value of $\vec{\mu}$, for example the agreement with the SM, one can define the profile likelihood ratio

$$\lambda(\vec{\mu}) = \frac{\mathcal{L}(\vec{\mu}, \hat{\hat{\vec{\theta}}}, \hat{\hat{\vec{\gamma}}})}{\mathcal{L}(\hat{\vec{\mu}}, \hat{\vec{\theta}}, \hat{\vec{\gamma}})} \tag{4.13}$$

where $\{\hat{\vec{\mu}}, \hat{\vec{\theta}}, \hat{\vec{\gamma}}\}$ are the values that maximise the likelihood overall and $\{\hat{\hat{\vec{\theta}}}, \hat{\hat{\vec{\gamma}}}\}$ the ones that maximise $\mathcal{L}$ for a fixed value of $\vec{\mu}$. The likelihood ratio is defined in the range [0,1], where a value close to 1 implies a good agreement between the data and the specific hypothesis. For larger disagreements, the likelihood ratio value moves towards values of 0.

From the likelihood ratio, a test statistic q_μ can be defined:

$$q_{\vec{\mu}} = \begin{cases} -2 \ln \lambda(\vec{\mu}) & \hat{\mu} \geq \mu \\ 0 & \hat{\mu} < \mu \end{cases} \tag{4.14}$$

In the case of $\mu = 0$, the hypothesis of no signal is tested, i.e. the null hypothesis. In this case, it is helpful to define the *p-value* to quantify the compatibility of the observed data with a background-only hypothesis ($\mu = 0$). The p-value is defined as:

$$p_{\bar{\mu}} = \int_{q_{\bar{\mu},\text{obs}}}^{\infty} f(q_{\bar{\mu}}|0) dq_{\bar{\mu}} \tag{4.15}$$

where $f(q_{\bar{\mu}}|0)$ is the probability density function of $q_{\bar{\mu}}$ under the assumption of $\mu = 0$ and $q_{\bar{\mu},\text{obs}}$ is the value of $q_{\bar{\mu}}$ observed in data. The p-value can be interpreted as the probability of finding data incompatible with the given hypothesis $\mu = 0$.

From the p-value, the number of Gaussian standard deviations Z above the background can be calculated as:

$$Z = \Phi^{-1}(1 - p), \tag{4.16}$$

where Φ^{-1} is the inverse Gaussian cumulative distribution function. Z is also known as the significance. A higher value of Z corresponds to a smaller p-value. In particular, a p-value of $\sim 3 \times 10^{-7}$ corresponds to a significance of $Z = 5$. In the accelerator-based particle physics community, this value is accepted as the threshold for the *discovery* of a process. The significance is usually quoted for processes where a reasonably large excess above the background-only hypothesis is expected—in the $VH(H \to c\bar{c})$ analysis this is the case for the two diboson signals.

For $VH(H \to c\bar{c})$, an upper limit at 95% confidence level (CL) on the $VH(H \to c\bar{c})$ signal strength is calculated since the significance is very small. The upper limit is determined using a modified frequentist CL_s method [90, 91]. For the calculation of the CL_s upper limit, the following test statistic is used:

$$q = -2 \ln \frac{\mathcal{L}_{s+b}}{\mathcal{L}_b} = -2 \ln \frac{\mathcal{L}(\mu = 1, \hat{\hat{\theta}}(1), \hat{\hat{\gamma}}(1))}{\mathcal{L}(\mu = 0, \hat{\hat{\theta}}(0), \hat{\hat{\gamma}}(0))} \tag{4.17}$$

where $\mathcal{L}_{s+b}$ is the likelihood of the nominal signal model and $\mathcal{L}_b$ is that of the background-only hypothesis. In particular, $\mathcal{L}_{s+b}$ corresponds to the likelihood with a signal strength of $\mu = 1$ and $\mathcal{L}_b$ corresponds to that with $\mu = 0$. A 95% CL limit can be calculated from this test statistic, which corresponds to the signal strength μ with a p-value of 0.05. The limit can be interpreted as the point where the probability of observing a dataset with a μ value above the determined limit is 5%.

In addition to performing a fit on the real dataset, expected results can be obtained by fitting an *Asimov* dataset, which corresponds to the sum of all simulated signal and background samples. In a *pre-fit* Asimov dataset, all nuisance parameters are set to their initial values (i.e. 0, or 1 for the floating normalisations), when constructing the dataset. In contrast, in a *post-fit* Asimov dataset, the nuisance parameters are set to their best-fit values from the fit to the real dataset. All expected results are obtained with a *post-fit* Asimov dataset unless stated otherwise.

4.8.2 Binning and Fit Model

The fit is performed simultaneously to the m_{cc} distributions in all signal and control regions in all three lepton channels. Different bin sizes are used in the various categories. The choice of binning is driven by the m_{cc} resolution of 10–20 GeV and the available data statistics. The bin sizes and fit ranges in the different regions are:

- Signal regions:
  - 16 bins from 50 to 210 GeV in the signal regions and *0-lepton* and *1-lepton* top control regions, with the exception of the *2-lepton*, 2 c-tag, $p_T^V > 150$ GeV signal regions.
  - 9 (10) bins from 50 to 185 (200) GeV in the *2-lepton*, 2 c-tag, $p_T^V > 150$ GeV, 2-jet (3-jet) signal regions.
- Control regions:
  - 13 bins from 80 to 340 GeV in each of the high-ΔR control regions, with the exception of the *2-lepton*, 2 c-tag, $p_T^V > 150$ GeV high-ΔR regions, where 9 bins from 80 to 350 GeV are used (i.e. twice the bin size of the signal regions)
  - A single bin from 50 to 210 GeV in each of the 0 c-tag control regions and *2-lepton* top control regions.

In summary, the following components enter the construction of the likelihood for the $VH(H \to c\bar{c})$ analysis:

- The signal strengths of the $VH(\to c\bar{c})$, $VZ(\to c\bar{c})$ and $VW(\to cq)$ processes: $\mu_{VH(c\bar{c})}$, $\mu_{VZ(c\bar{c})}$ and $\mu_{VW(cq)}$.
- The m_{cc} distributions of all of the signal and backgrounds, as predicted by simulation
- Systematic uncertainties as nuisance parameters
  - The experimental uncertainties related to the detector performance described in Sect. 4.6. These includes uncertainties related to the identification and reconstruction of jets, leptons, $E_{\mathrm{T}}^{\mathrm{miss}}$, flavour tagging and truth tagging.
  - Floating normalisation parameters for all of the major backgrounds (V+jets and top quark backgrounds).
  - Modelling uncertainties derived using alternative generators and scale variations: Normalisation uncertainties, acceptance ratios, flavour compositions and m_{cc} shape uncertainties. A detailed description is provided in Sect. 4.7.
  - Uncertainties on the statistics of the simulated samples.

4.9 Results

4.9.1 Signal Strength Results

The best fit signal strengths for the $VH(H \to c\bar{c})$, $VZ(Z \to c\bar{c})$ and $VW(W \to cq)$ processes are:

$$\begin{aligned}\mu_{VH(c\bar{c})} &= -9 \pm 10\,(\text{stat.}) \pm 12\,(\text{syst.})\\ \mu_{VW(cq)} &= 0.83 \pm 0.11\,(\text{stat.}) \pm 0.21\,(\text{syst.})\\ \mu_{VZ(c\bar{c})} &= 1.16 \pm 0.32\,(\text{stat.}) \pm 0.36\,(\text{syst.}).\end{aligned}$$

The anti-correlation between the $VH(\to c\bar{c})$ signal strength and $VW(\to cq)$ signal strength is 17%, and the anti-correlation with the $VZ(\to c\bar{c})$ signal strength is 16%. The probability of compatibility with the SM, i.e. all three signal strengths being equal to unity, is 84%. The floating normalisation parameters of the main backgrounds are shown in Table 4.7. Generally, the normalisations of the top quark backgrounds are slightly higher than the prediction of the nominal $t\bar{t}$ and Wt MC samples. For the V+jets background, the heavy flavour and mixed flavour normalisations are slightly higher than the predictions of the nominal simulated samples, and the normalisations of the light flavour components are close to the simulation predictions. Overall, all normalisations are approximately within one standard deviation of their pre-fit value.

Figure 4.25 shows the m_{cc} distributions after the fit, with all of the backgrounds subtracted. The only remaining contributions are the $VH(H \to c\bar{c})$, $VW(W \to cq)$ and $VZ(Z \to c\bar{c})$ signals. In particular, the diboson signals are clearly visible. The $VW(W \to cq)$ signal dominates in the 1 c-tag category, while the $VZ(Z \to c\bar{c})$ signal is more important in the 2 c-tag category. The uncertainty of the background is shown as the dashed area. The expected and observed significances of the diboson processes can be determined from the fit. For the $VW(W \to cq)$ process, a significance of 3.8 σ is observed, for 4.6 σ expected. For the $VZ(Z \to c\bar{c})$ process, a significance of 2.6 σ is observed, for 2.2 σ expected. The best fit $VH(H \to c\bar{c})$ signal is shown in red, and a downwards fluctuation of the data can be seen for m_{cc} around 125 GeV, leading to a negative signal strength.

Figure 4.26 (left) shows the $VH(H \to c\bar{c})$ signal strengths for the combined fit and in each channel individually. The single-channel values are determined by fitting all analysis categories but using different signal strength parameters for each channel. The signal strengths observed in each channel agree well, and the compatibility with the combined fit is 95%. A 95% CL limit on the $VH(H \to c\bar{c})$ process can also be determined. Figure 4.26 (right) shows the resulting limits in the individual channels and the combined fit. A limit of 26 times the SM prediction is observed for the combined fit, for an expected limit of 31 times the SM prediction. Of note is that the *0-lepton* channel is the most sensitive, followed by the *2-lepton* and then the *1-lepton* channel.

Table 4.7 **Values of the free-floating background normalisation parameters obtained from the likelihood fit to data. The uncertainties represent the combined statistical and systematic uncertainties.** Unless otherwise stated, normalisation parameters are correlated across all p_T^V and number-of-jets analysis regions

Background	p_T^V	Jets	Value
Top(b)			0.91 ± 0.06
Top(other)			0.94 ± 0.08
$t\bar{t}$ (*2-lepton*)	$p_T^V > 150$ GeV	2	0.76 ± 0.22
		3	0.96 ± 0.13
	$75 < p_T^V < 150$ GeV	2	1.08 ± 0.08
		3	1.06 ± 0.07
$W + hf$			1.16 ± 0.35
$W + mf$			1.28 ± 0.14
$W + lf$		2	1.02 ± 0.04
		3	0.97 ± 0.05
$Z + hf$	$p_T^V > 150$ GeV		1.19 ± 0.22
	$75 < p_T^V < 150$ GeV		1.25 ± 0.25
$Z + mf$	$p_T^V > 150$ GeV		1.10 ± 0.15
	$75 < p_T^V < 150$ GeV		1.11 ± 0.15
$Z + lf$	$p_T^V > 150$ GeV	2	1.07 ± 0.03
		3	1.08 ± 0.05
	$75 < p_T^V < 150$ GeV	2	1.12 ± 0.04
		3	1.07 ± 0.06

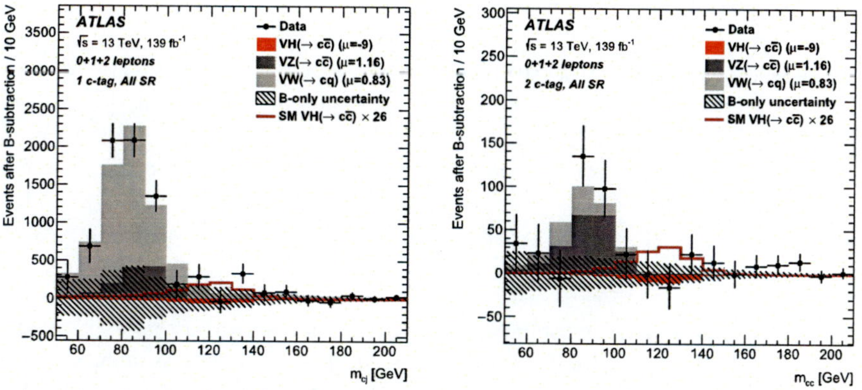

Fig. 4.25 The post-fit m_{cc} distribution summed over all signal regions after subtracting backgrounds, leaving only the $VH(H \to c\bar{c})$, $VW(W \to cq)$ and $VZ(Z \to c\bar{c})$ processes, for events with one c-tag (left) and two c-tags (right). The red filled histogram corresponds to the $VH(H \to c\bar{c})$ signal for the fitted value of $\mu_{VH(c\bar{c})} = -9$, while the open red histogram corresponds to the signal expected at the 95% CL upper limit on $\mu_{VH(c\bar{c})}$ ($\mu_{VH(c\bar{c})} = 26$). The hatched band shows the uncertainty of the fitted background

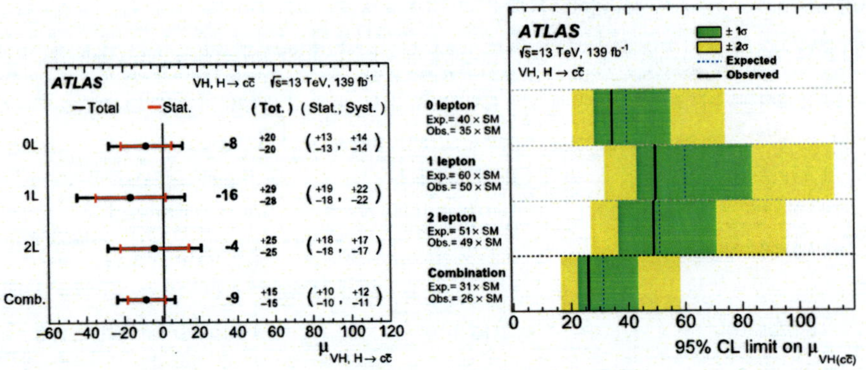

Fig. 4.26 The observed signal strengths (left), and expected and observed 95% CL upper limits on the cross-section times branching fraction (right) in each lepton channel and for the combined fit. The single-channel values are obtained using a five-POI fit, in which each channel has a separate $VH(H \to c\bar{c})$ parameter of interest, in addition to the two diboson POIs

This result significantly improves on the previous iteration of the analysis, where a limit of 110 times the SM prediction was observed (for 150 × SM expected). The main improvements are the addition of the *0-lepton* and *1-lepton* channels and the increase of the dataset from 36 fb^{-1} to 139 fb^{-1}. Improvements on the expected limit are also due to the new flavour tagging working point (+36%), the additional splits of the signal regions (+10%) and the addition of control regions (+10%).

Breakdown of Uncertainties

Table 4.8 shows the impact of the different groups of NPs on the signal strengths μ of $VH(H \to c\bar{c})$, $VW(W \to cq)$ and $VZ(Z \to c\bar{c})$. The impact of a group of NPs is obtained from a fit where all NPs belonging to that particular group are fixed to their post-fit values. This fit then results in a new best-fit value $\hat{\mu}'$ and uncertainty $\sigma_{\hat{\mu}'}$. The computed uncertainty $\sigma_{\hat{\mu}'}$ is then compared to nominal one $\sigma_{\hat{\mu}}$ and the impact is defined as:

$$\text{impact} = \sqrt{\sigma_{\hat{\mu}}^2 - \sigma_{\hat{\mu}'}^2}. \tag{4.18}$$

The impact of the statistical uncertainties is defined as the uncertainty of a fit with all NPs fixed, except for the floating normalisations. The total systematical uncertainty is calculated as the quadratic difference of the statistical and the total uncertainty. The sum in quadrature of the individual contributions differs from the total uncertainty due to correlations between the NPs.

For the $VH(H \to c\bar{c})$ signal strength, the statistical and systematic uncertainties are of similar size, with the systematic uncertainties being slightly larger. In terms of the systematic uncertainties, the largest contribution comes from the modelling of the background processes, in particular Z+jets and the top quark backgrounds. The Z+jets background plays a significant role in the *0-lepton* and *2-lepton* channels.

Table 4.8 Breakdown of contributions to the uncertainty in the fitted values of $\mu_{VH(c\bar{c})}$, $\mu_{VW(cq)}$ and $\mu_{VZ(c\bar{c})}$. The sum in quadrature of uncertainties from different sources may differ from the total due to correlations. In cases where the upward and downward systematic variations have different values, the mean of the absolute values is shown

Source of uncertainty		$\mu_{VH(c\bar{c})}$	$\mu_{VW(cq)}$	$\mu_{VZ(c\bar{c})}$
Total		15.3	0.24	0.48
Statistical		10.0	0.11	0.32
Systematic		11.5	0.21	0.36
Statistical uncertainties				
Data sample size only		7.8	0.05	0.23
Floating normalisations		5.1	0.09	0.22
Theoretical and modelling uncertainties				
$VH(H \to c\bar{c})$		2.1	<0.01	0.01
Z+jets		7.0	0.05	0.17
Top quark		3.9	0.13	0.09
W+jets		3.0	0.05	0.11
Diboson		1.0	0.09	0.12
$VH(H \to b\bar{b})$		0.8	<0.01	0.01
Multi-jet		1.0	0.03	0.02
Simulation samples size		4.2	0.09	0.13
Experimental uncertainties				
Jets		2.8	0.06	0.13
Leptons		0.5	0.01	0.01
$E_{\mathrm{T}}^{\mathrm{miss}}$		0.2	0.01	0.01
Pile-up and luminosity		0.3	0.01	0.01
Flavour tagging	c-jets	1.6	0.05	0.16
	b-jets	1.1	0.01	0.03
	light-jets	0.4	0.01	0.06
	τ-jets	0.3	0.01	0.04
Truth-flavour tagging	ΔR correction	3.3	0.03	0.10
	Residual non-closure	1.7	0.03	0.10

Since the distribution of the Top(b) background peaks close to the Higgs mass at around 120 GeV, it can also have a sizeable effect. In terms of experimental uncertainties, the most considerable contributions come from uncertainties related to jet reconstruction and calibration, and flavour tagging. The simulation sample uncertainties and the uncertainties related to truth-tagging also yield significant contributions.

For the $VW(W \to cq)$ signal strength, the systematic uncertainties are twice as large as the statistical ones. The modelling of the top quark background gives is the largest contribution to the systematic uncertainties. This is mainly due to the Top(other) component, which is peaked at the W-boson mass, where the $VW(W \to cq)$ peak is also located. The modelling uncertainties on the diboson processes themselves also have a sizeable impact. Out of the experimental uncertainties, the uncertainties related to jet reconstruction and flavour tagging have a significant contribution.

For the $VZ(Z \to c\bar{c})$ process, the breakdown of uncertainties shows a similar pattern as the one for $VH(H \to c\bar{c})$. The statistical and systematic uncertainties are of similar size. The largest contributions to the systematic uncertainties come from the modelling of the V+jets, diboson and top quark processes. The experimental uncertainties have equally large effects, with the uncertainties on the flavour tagging of c-jets, jet reconstruction and truth-tagging being the largest.

Ranking of Nuisance Parameters

The impact of individual NPs on the signal strength uncertainties can also be determined to rank the most important single NPs in the analysis. This can be done by changing the values of the NPs upwards and downwards by their post-fit uncertainties and repeating the fit. The resulting difference of the best fit value with respect to the nominal one can then be calculated and used to rank the NPs.

The ranking of nuisance parameters for the $VH(H \to c\bar{c})$ signal strength is shown in Fig. 4.27. The highest-ranked nuisance parameters are one of the Z+jets m_{cc} shape uncertainties and the extrapolation uncertainty of the top quark background from the control region into the signal region. Both of these uncertainties are related to extrapolating from a control region, so they are applied in the signal region only, which explains why they can have such a large impact. The signal strengths of the diboson processes are also highly ranked due to their sizeable correlations with the $VH(H \to c\bar{c})$ signal strength. Other highly ranked uncertainties include the truth-tagging ΔR correction uncertainties, further V+jets m_{cc} shape uncertainties and one simulation statistics nuisance parameter, corresponding to a particular bin at $m_{cc} =$ 125 GeV. Overall, most of the highly ranked nuisance parameters modify the shape of the m_{cc} distribution in the signal region. Their high rank can be explained by the small size of the signal and the fact that normalisation uncertainties are generally well constrained from the control regions for all of the major backgrounds.

4.9.2 Constraints on the κ_c Coupling Modifier

Beyond placing a simple limit on the $VH(H \to c\bar{c})$ process, it is also possible to directly probe the Higgs coupling to charm quarks. This can be done within the κ framework [61, 92], which introduces the Higgs coupling modifiers κ_i, defined as the deviation of the Higgs Yukawa coupling from the Standard Model expectation, for example, $\kappa_c = \frac{y_c}{y_c^{SM}}$. Figure 4.28 shows the instances in the $VH(H \to c\bar{c})$ decay, where the coupling modifiers can have an impact. This is the case in the production in association with the W or Z boson, or directly in the Higgs decay into the two

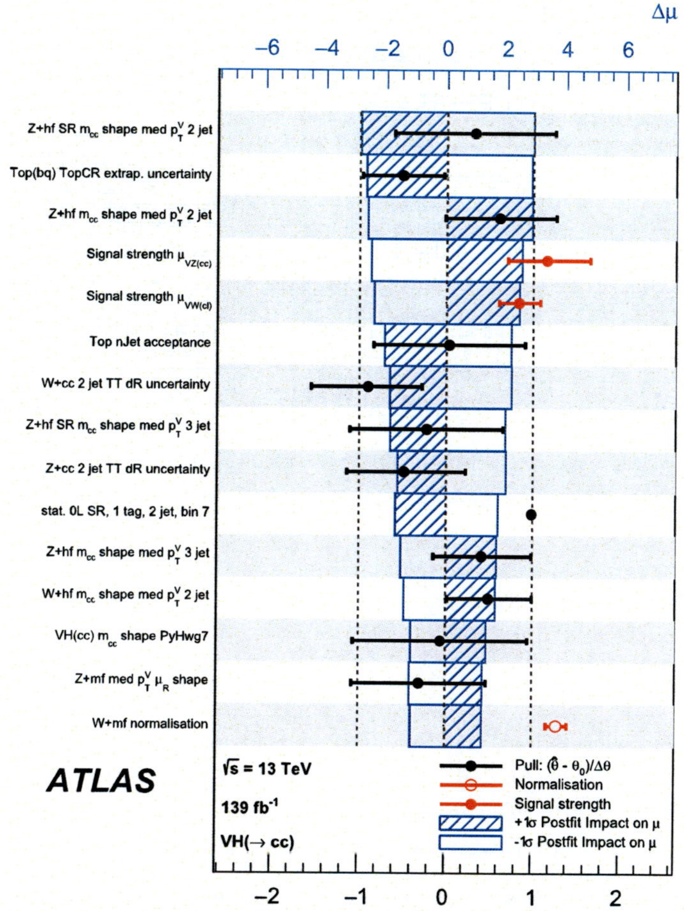

Fig. 4.27 Impact of systematic uncertainties for the 15 largest contributions on the fitted $VH(H \to c\bar{c})$ signal-strength parameter $\hat{\mu}$ sorted in decreasing order. The boxes show the variations of $\hat{\mu}$, referring to the top x-axis, when fixing the corresponding individual nuisance parameter to its post-fit value modified upwards or downwards by its post-fit uncertainty, i.e. $\hat{\theta} \pm \sigma_{\hat{\theta}}$, and repeating the fit. The impact of up- and down-variations can be distinguished via the dashed and plane box fillings. The filled circles show the deviation of the fitted value for each nuisance parameter, $\hat{\theta}$, from their nominal input value θ_0 expressed in standard deviations with respect to their nominal uncertainties $\Delta\theta$ (bottom x-axis). The error bars show the post-fit uncertainties on $\hat{\theta}$ with respect to their nominal uncertainties

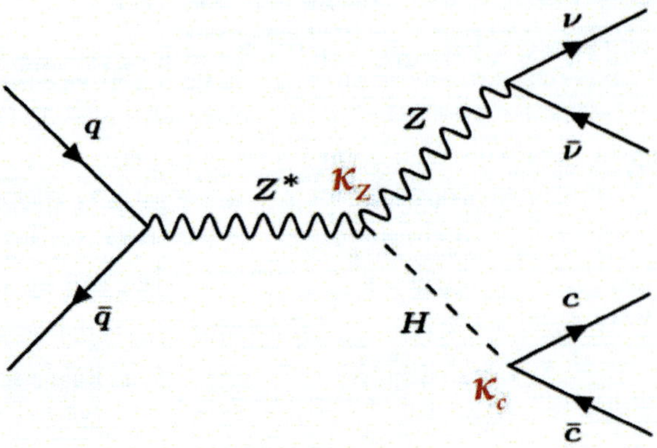

Fig. 4.28 Feynman diagram of the $VH(H \to c\bar{c})$ process and points where κ coupling modifiers could modify the Higgs coupling

c-quarks. The effect of the modified couplings can then be seen in the measured cross-section times branching fraction, where κ modifiers enter as follows:

$$\sigma \times BR(VH(\to c\bar{c})) = \sigma_{SM} \times BR(VH(\to c\bar{c}))_{SM} \times \kappa_V^2 \times \frac{\kappa_c^2}{\kappa_H^2} \tag{4.19}$$

In the case of VH production, just the κ_V modifier is included in the equation. In the case of the $H \to c\bar{c}$ decay, the modification to the total Higgs decay width also has to be considered. This modification is parametrised by the parameter κ_H, a sum of all the Standard Model branching fractions and their coupling modifiers, and any potential BSM Higgs decays and can be schematically written as:

$$\kappa_H^2 \sim 0.57\kappa_b^2 + 0.22\kappa_W^2 + 0.03\kappa_c^2 + \cdots + \text{BSM decays} \tag{4.20}$$

Several assumptions are made to simplify the above equations in the interpretation of the $VH(H \to c\bar{c})$ search in terms of the κ coupling modifiers. Only the κ_c modifier is considered, i.e. only the direct Higgs decays are taken into account, and all other coupling modifiers are set to 1. Additionally, the assumption is made that there is no additional contribution from BSM decays to the total Higgs width. With these assumptions, the $VH(H \to c\bar{c})$ signal strength can be parameterised as

$$\mu_{VH(c\bar{c})}(\kappa_c) = \frac{\kappa_c^2}{1 + B^{\text{SM}}_{H\to c\bar{c}}(\kappa_c^2 - 1)}, \tag{4.21}$$

where $B^{\text{SM}}_{H\to c\bar{c}}$ is the $H \to c\bar{c}$ branching fraction predicted in the SM. The resulting signal strength as a function of κ_c is shown in Fig. 4.29. With large values of κ_c, the parametrisation asymptotically approaches $1/B^{\text{SM}}_{H\to c\bar{c}}$. As a consequence, using this

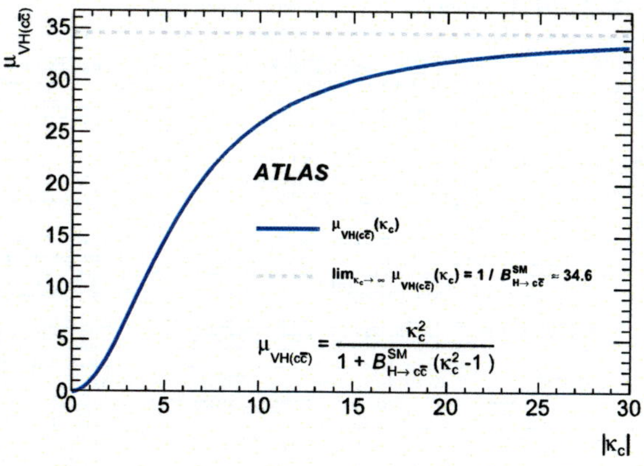

Fig. 4.29 The parametrisation of $\mu_{VH(c\bar{c})}$ versus κ_c. As $\kappa_c \to \infty$, $\mu_{VH(c\bar{c})}$ approaches $1/B^{SM}_{H\to c\bar{c}}$

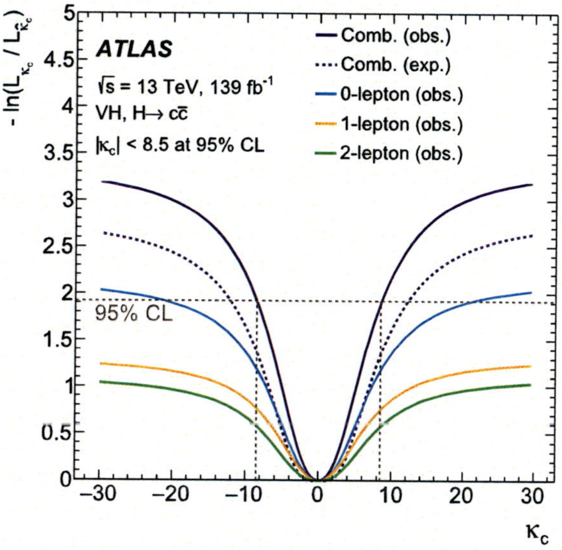

Fig. 4.30 Expected and observed profile likelihood scans for κ_c. The single-channel likelihoods are obtained using a five-POI fit, in which each channel has a separate $VH(H \to c\bar{c})$ parameter of interest, in addition to the two diboson POIs. The order of the lines in the plot matches the order of the lines in the legend

parametrisation, the analysis is only sensitive to κ_c if the limit on the $VH(H \to c\bar{c})$ signal strength is less than 35 times the SM prediction.

Using this parametrisation, it is possible to perform a fit to the κ_c parameter directly, using the same setup as for the previous results, but replacing the $\mu_{VH(c\bar{c})}$ parameter with κ_c. The resulting scans of the likelihood as a function of κ_c are shown in Fig. 4.30 for the individual channels and the combined fit. In the combined fit an observed limit of $|\kappa_c| < 8.5$ is found at the 95% CL, for an expected limit of $|\kappa_c| < 12.4$. This is the first time a direct limit on κ_c is set from direct Higgs decays into charm quarks.

4.10 Statistical Combination of the $VH(H \to c\bar{c})$ and $VH(H \to b\bar{b})$ Analyses

The flavour tagging strategy of the $VH(H \to c\bar{c})$ analysis was designed to ensure no significant overlap with the existing $VH(H \to b\bar{b})$ analysis. This allows for the two analyses to be combined straightforwardly. There is always some $VH(H \to c\bar{c})$ signal in the $VH(H \to b\bar{b})$ regions and vice versa, so it is possible to fit the two signal strengths simultaneously in all categories of the $VH(H \to b\bar{b})$ and $VH(H \to c\bar{c})$ analyses. In addition to the small additional signal from the additional fit regions, this combination makes it possible to constrain the ratio κ_c/κ_b. Fitting the ratio of coupling modifiers relaxes some of the assumptions previously made in the κ_c interpretation of the $VH(H \to c\bar{c})$ analysis and provides a simultaneous measurement of the Higgs coupling to b- and c-quarks.

4.10.1 *The* $VH(H \to b\bar{b})$ *Measurement*

The $VH(H \to b\bar{b})$ analysis is very similar to the $VH(H \to c\bar{c})$ analysis in many aspects. It targets the measurement of the Higgs decay into b-quarks, produced in association with a W or Z boson. The most recent analysis utilises the entire Run 2 dataset at 13 TeV to measure the $VH(H \to b\bar{b})$ process, with the details described in [77]. The measurement was performed separately in the WH and ZH production modes, as well as in cross-section measurements in the reduced VH stage-1.2 simplified template cross-section (STXS) scheme [93, 94], where STXS bins are defined based on p_T^V and jet multiplicity.

As in the $VH(H \to c\bar{c})$ analysis, events are categorised into a *0-lepton*, *1-lepton* and *2-lepton* channel based on the decay of the vector boson. Events are further categories by jet multiplicity, i.e. 2 or 3+ jets, and p_T^V, with three possible p_T^V categories ($75 < p_T^V < 150$ GeV (*2-lepton* channel only), $150 < p_T^V < 250$ GeV and $p_T^V > 250$ GeV). The signal regions are defined through flavour tagging by requiring exactly two b-tags. The b-jets are identified using the `MV2c10` algorithm at a 70% efficiency working point [78]. The selection leads to 14 signal regions, each of which has a corresponding low and high ΔR control region to constrain the backgrounds in the analysis. Similar to the $VH(H \to c\bar{c})$ analysis, the main backgrounds in the analysis come from the V+jets and $t\bar{t}$ processes. In the $VH(H \to b\bar{b})$ analysis regime, these backgrounds are enriched in b-jets.

Instead of the invariant mass, a multivariate discriminant is used as the discriminating variable to improve the sensitivity of the analysis. A boosted decision tree (BDT) is trained in each channel. The BDT discriminant is used to perform a fit to the signal regions and yield-only control regions simultaneously. A value of $\mu_{VH(b\bar{b})} = 1.02^{+0.18}_{-0.17}$ was observed for the $VH(H \to b\bar{b})$ signal strength, which corresponds to a significance of 6.7σ.

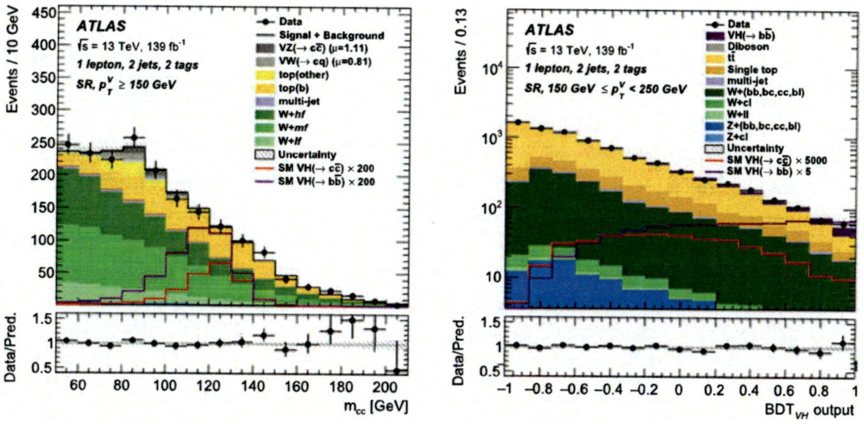

Fig. 4.31 Examples of the fitted distributions in the $VH(H \to c\bar{c})$ (left) and $VH(H \to b\bar{b})$ (right) categories in the *1-lepton* channel

4.10.2 Combination Setup

Before performing the statistical combination, it is important to understand the contamination of $VH(H \to b\bar{b})$ events in the $VH(H \to c\bar{c})$ analysis and vice versa. Figure 4.31 shows examples of the fitted distributions in the *1-lepton* channels for the $VH(H \to b\bar{b})$ and $VH(H \to c\bar{c})$ regions respectively. For the $VH(H \to c\bar{c})$ region, the m_{cc} discriminant is shown, and the two signals are shown as overlays, scaled by a factor of 200. Depending on the category, the $VH(H \to b\bar{b})$ signal is approximately 2–7 times the size of the $VH(H \to c\bar{c})$ signal due to the significantly larger $H \to b\bar{b}$ branching ratio. However, the $VH(H \to b\bar{b})$ contribution is still small compared to the total background. For the $VH(H \to b\bar{b})$ analysis, the BDT discriminant is shown, and the $VH(H \to c\bar{c})$ signal is scaled by a factor 1000 compared to the $VH(H \to b\bar{b})$ one. In the $VH(H \to b\bar{b})$ categories, the $VH(H \to c\bar{c})$ signal is of the order of 0.15% of the $VH(H \to b\bar{b})$ signal and negligible compared to the total background. The added additional sensitivity from the combination is small for both signals.

In the combination, a likelihood fit is performed simultaneously to all of the $VH(H \to b\bar{b})$ and $VH(H \to c\bar{c})$ analysis regions. The parameters of interest are the signal strengths $\mu_{VH(c\bar{c})}$ and $\mu_{VH(b\bar{b})}$, which are free to float in all of the analysis categories. Additionally, the diboson signal strengths, $\mu_{VZ(c\bar{c})}$ and $\mu_{VW(cq)}$ are also left to float freely in the $VH(H \to c\bar{c})$ analysis categories. The fit model is kept as close as possible to the standalone analyses. The only change in the $VH(H \to b\bar{b})$ regions is the addition of the $VH(H \to c\bar{c})$ signal template, which was previously neglected. In the $VH(H \to c\bar{c})$ categories, the $VH(H \to b\bar{b})$ background was already considered. In the combination, it is now treated as a signal, and the overall normalisation uncertainty is removed. Experimental systematic uncertainties, common to both analyses, are correlated between the two analysis regimes. The only exception is the flavour tagging, where the b- and c-tagging use different algorithms and calibrations and thus can not be correlated. Background normalisations and modelling uncertainties are considered uncorrelated between the two analyses. The modelling uncertainties are derived in distinct ways due to the different flavour compositions and the use of a BDT in the $VH(H \to b\bar{b})$ analysis. The floating normalisations are also kept uncorrelated in both analyses. A test was performed where normalisations were correlated, and the resulting impact was minor.

Initially, a fit to $\mu_{VH(c\bar{c})}$ and $\mu_{VH(b\bar{b})}$ is performed, but it is again possible to perform the analysis to measure the κ coupling modifiers. With now both κ_c and κ_b free to float, the parametrisation of the signal strengths can be extended as:

$$\mu_{VH(b\bar{b})} = \frac{\kappa_b^2}{1 - B(H \to b\bar{b}) - B(H \to c\bar{c}) + B(H \to c\bar{c})\kappa_c^2 + B(H \to b\bar{b})\kappa_b^2} \tag{4.22}$$

$$\mu_{VH(c\bar{c})} = \frac{\kappa_c^2}{1 - B(H \to b\bar{b}) - B(H \to c\bar{c}) + B(H \to c\bar{c})\kappa_c^2 + B(H \to b\bar{b})\kappa_b^2} \tag{4.23}$$

In the case of the $VH(H \to b\bar{b})$ signal, contributions of κ_b in the ggH production loop are also considered [95], while for the $VH(H \to c\bar{c})$ signal no such parametrisation is available. In both cases, the impact is negligible, and the contribution is not shown in the above parametrisation. When taking the ratio of κ_c/κ_b, the denominator in the two equations cancels, which means that it is not necessary to make assumptions on the total Higgs width. A fit is therefore performed to constrain κ_c/κ_b, while κ_b is also allowed to float freely.

4.10.3 $VH(H \to b\bar{b}/c\bar{c})$ Combination Results

A fit to the $VH(H \to c\bar{c})$ and $VH(H \to b\bar{b})$ signal strengths is performed. The resulting signal strengths and the results from the standalone analyses are shown in Table 4.9. The likelihood scan of $\mu_{VH(b\bar{b})}$ and $\mu_{VH(c\bar{c})}$ for the expected and observed results is shown in Fig. 4.32. The best-fit values agree very well with the individual analyses. As expected, there is no significant reduction of uncertainties, as the uncertainties are left largely uncorrelated between the two analyses.

Further tests were performed on these results. While the signal regions of the two analyses are fully orthogonal, there can be a small number (less than 5%) of overlapping events in the $VH(H \to c\bar{c})$ top control region in the *0-lepton* and *1-lepton* channel and the 0 c-tag control region in the *1-lepton* and *2-lepton* channel. A test was performed by removing the overlapping events, and no impact on the resulting best fit values or uncertainties was found. Additionally, decorrelating all experimental uncertainties was tested, as well as correlating some of the normalisation uncertainties. In either case, the impact was negligible.

Table 4.9 Best fit values for $\mu_{VH(b\bar{b})}$, $\mu_{VH(c\bar{c})}$, $\mu_{VZ(c\bar{c})}$ and $\mu_{VW(cq)}$ in the individual $VH(H \to c\bar{c})$ and $VH(H \to b\bar{b})$ analyses and in the combination

	$\mu_{VH(b\bar{b})}$	$\mu_{VH(c\bar{c})}$	$\mu_{VW(cq)}$	$\mu_{VZ(c\bar{c})}$
$VH(H \to b\bar{b})$	1.02 ± 0.18	–	–	–
$VH(H \to c\bar{c})$	–	-8.6 ± 15.3	0.83 ± 0.24	1.16 ± 0.48
$VH(H \to b\bar{b}/c\bar{c})$ combination	1.06 ± 0.18	-9.1 ± 15.3	0.81 ± 0.23	1.11 ± 0.46

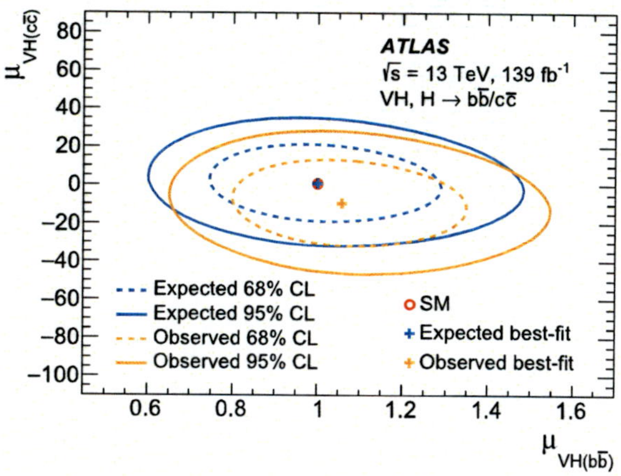

Fig. 4.32 The observed and expected best fit values of $\mu_{VH(c\bar{c})}$ and $\mu_{VH(b\bar{b})}$ with their 68 and 95% CL contours

Then, a fit to the κ coupling modifiers can be performed. The resulting likelihood scan of κ_b and κ_c is shown in Fig. 4.33. Due to the inclusion of κ_b in the parametrisation of the ggH production loop, the likelihood is slightly asymmetric in κ_b (but not κ_c), which results in two expected minima. The observed best fit value is found to be $(\kappa_b, \kappa_c) = (-1.02, 0)$.

When fitting the ratio $|\kappa_c/\kappa_b|$, without any assumptions on the Higgs width, the observed limit at 95% CL is $|\kappa_c/\kappa_b| < 4.5$ (for $|\kappa_c/\kappa_b| < 5.1$ expected). This is smaller than the ratio of the b- and c-quark masses, $m_b/m_c = 4.578 \pm 0.008$ [96]. Therefore, this measurement constrains the coupling of the Higgs boson to the charm quark to be weaker than the coupling of the Higgs boson to the bottom quark (at 95% CL). A likelihood scan of $|\kappa_c/\kappa_b|$ is shown in Fig. 4.34, with the point where the Higgs coupling to b- and c-quarks would be equal shown as a green line.

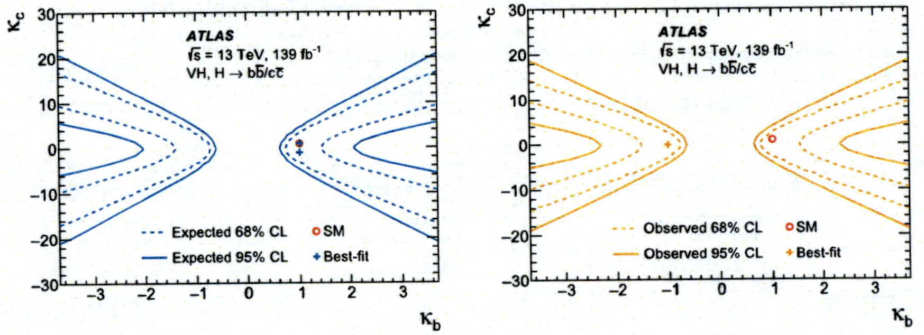

Fig. 4.33 The expected (left) and observed (right) constraints on κ_c and κ_b at 68 and 95% confidence levels. The grey lines correspond to the values of κ_c and κ_b for which the Higgs-charm and Higgs-bottom couplings are equal, where each coupling strength $|\kappa_i y_i|$ is the product of the κ_i modifier and the Yukawa coupling, y_i, for $i = b, c$

Fig. 4.34 Expected and observed κ_c/κ_b profile likelihood scans of the combined $VH(H \to c\bar{c})$ and $VH(H \to b\bar{b})$ likelihood function. The vertical green lines correspond to the values of $|\kappa_c/\kappa_b|$ for which the Higgs-charm and Higgs-bottom couplings are equal, where each coupling strength $|\kappa_i y_i|$ is the product of the κ_i modifier and the Yukawa coupling, y_i, for $i = b, c$, and is equal to $m_b/m_c = 4.578 \pm 0.008$ [96]

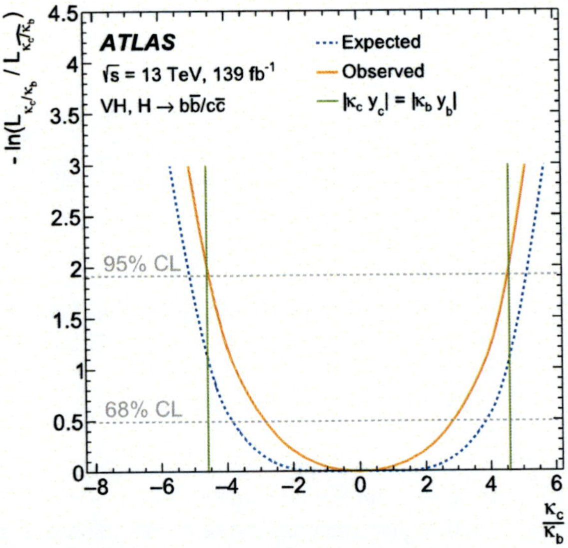

4.11 Extrapolation of the $VH(H \to c\bar{c})$ Analysis to HL-LHC

Around 2028, the LHC is expected to upgrade the accelerator complex and its detectors to operate at a higher instantaneous luminosity and a larger number of collisions per bunch-crossing. The High Luminosity LHC (HL-LHC) is expected to collect a dataset of 3000 fb^{-1} at a center-of-mass energy of 14 TeV. An extrapolation study was performed to understand the expected sensitivity of the $VH(H \to c\bar{c})$ analysis with the dataset collected at HL-LHC, based on the Run 2 analysis. This study includes the extrapolation of the $VH(H \to b\bar{b})$ analysis, the $VH(H \to c\bar{c})$ analysis and the $VH(H \to b\bar{b}/c\bar{c})$ combination. The details of the extrapolation analysis can be found in [97], and the $VH(H \to c\bar{c})$ and $VH(H \to b\bar{b}/c\bar{c})$ combination results are summarised below. The extrapolation strategy follows the analysis strategy of the published analyses. It also quantifies the impact of reduced systematic uncertainties and improvements in the flavour tagging performance in the $VH(H \to c\bar{c})$ analysis.

4.11.1 Extrapolation Setup

The extrapolation of the analyses to the expectation for the HL-LHC is performed by scaling the signal and background expectations to an increased integrated luminosity and center-of-mass energy. First, a scale factor is applied to the normalisations of the signal and background predictions for all processes in all categories, which accounts for the increase in integrated luminosity from 139 to 3000 fb^{-1}. Next, scale factors are applied to the different processes to account for the increase in center-of-mass energy from 13 to 14 TeV, which are derived using the expected cross-sections [92]. Depending on the process, these scale factors vary between 1.10 and 1.18.

Experimental and theoretical uncertainties are included in the extrapolation, based on those used in the Run 2 analyses. However, the size of these uncertainties is scaled to account for possible reductions due to the larger available dataset, for example, in improved calibrations of physics objects [98, 99]. A summary of the scale factors applied to the systematic uncertainties is shown in Table 4.10. Experimental uncertainties related to $E_{\text{T}}^{\text{miss}}$ are scaled by 1/2, and the luminosity uncertainty is reduced to 1% (0.58 of the value in Run 2). Flavour tagging uncertainties are scaled by a factor of 1/2, except for those related to light-jets in the $VH(H \to c\bar{c})$ analysis, which are left unscaled. All other experimental uncertainties are kept at their Run 2 levels, and their impact on the results is small. Theory and background uncertainties on the signal and background processes are also scaled by a factor of 1/2. Uncertainties related to the statistics of the simulated samples are neglected, which also includes uncertainties associated with the *truth-tagging* of jets. It is expected that the ATLAS simulation and reconstruction software will improve in the future, such that simulation statistics uncertainties will become negligible.

The extrapolation analyses are performed to determine the expected values of the quantities measured in the Run 2 analyses at HL-LHC. In particular, for the extrapolation of the $VH(H \to c\bar{c})$ analysis and the $VH(H \to b\bar{b}/c\bar{c})$ combination this means:

$VH(H \to c\bar{c})$ fit setup For the extrapolation of the $VH(H \to c\bar{c})$ search, a fit is performed on the simulated dataset, determining the expected signal strength of the $VH(H \to c\bar{c})$ process, while the $VZ(Z \to c\bar{c})$ and $VW(W \to cq)$ signal strengths are free to float. Additionally, an upper limit on the $VH(H \to c\bar{c})$ signal strength is determined at 95% CL, and the expected constraint on κ_c is computed.

Table 4.10 Summary of scale factors applied to systematic uncertainties in the extrapolation. The luminosity uncertainty is reduced from 1.7 to 1.0%, resulting in a scale factor of 0.58

Uncertainties	Scale factor
E_T^{miss}	0.5
Lepton	1
Jet	1
Flavour tagging c-, b- and τ-jets	0.5
Flavour tagging light-jets (MV2c10 in $VH(H \to b\bar{b})$)	0.5
Flavour tagging light-jets (DL1 in $VH(H \to c\bar{c})$)	1.0
Luminosity	0.58
Signal modelling	0.5
Background modelling	0.5
MC statistics	0
Truth-tagging uncertainties ($VH(H \to c\bar{c})$ only)	0

$VH(H \to c\bar{c})$ **combination fit setup** Following the methodology described in Sect. 4.10 a simultaneous fit to all $VH(H \to b\bar{b})$ and $VH(H \to c\bar{c})$ regions is performed. First, the expected signal strengths for $VH(H \to b\bar{b})$ and $VH(H \to c\bar{c})$ are determined. Then, a fit to the κ coupling modifiers is performed, first measuring κ_b and κ_c simultaneously. Finally, a fit to $|\kappa_c/\kappa_b|$ is performed, with κ_b being allowed to float in the fit.

4.11.2 Extrapolation Results of the $VH(H \to c\bar{c})$ Analysis

The results of a fit to the $VH(H \to c\bar{c})$ signal strength, extrapolated to the HL-LHC, are shown in Fig. 4.35, along with the corresponding 95% CL limits. Results are shown for separate fits in which parameters of interest for the individual lepton channels are either separated or combined. In the combined fit, the best fit value is:

$$\mu_{VH(c\bar{c})} = 1.0 \pm 3.2 = 1.0 \pm 2.0(\text{stat.})^{+2.6}_{-2.5}(\text{syst.})$$

The expected upper limit on $\mu_{VH(c\bar{c})}$ is found to be 6.4 times the SM prediction at 95% CL. Table 4.11 shows the contribution of different groups of uncertainties. The statistical and systematic uncertainties are of similar size. The largest single contribution to systematic uncertainties comes from modelling the Z+jets background, which agrees with what was observed for the $VH(H \to c\bar{c})$ Run 2 analysis. Similarly, out of the experimental uncertainties, those related to the flavour tagging still have the largest impact.

Several studies were performed to quantify the impact of different groups of uncertainties. The scaling of the signal and background uncertainties has a relative

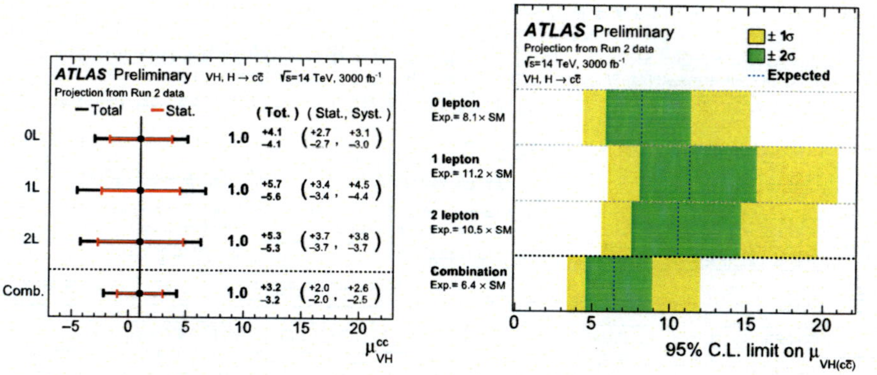

Fig. 4.35 Best fit signal strengths (left) and limits (right) on the $VH(H \to c\bar{c})$ signal strength in the different lepton channels from a fit to an Asimov dataset scaled to 3000 fb^{-1} and $\sqrt{s} = 14$ TeV. The per-channel results are obtained using a five-POI fit, in which each channel has a separate $VH(H \to c\bar{c})$ parameter of interest

effect of around 10% when decreasing or increasing the systematic uncertainty scale factors by a factor 2. Due to the relatively low c-tagging efficiency, statistical uncertainties related to the available simulated sample size have a non-negligible impact in the Run 2 $VH(H \to c\bar{c})$ analysis. While these uncertainties were neglected for the baseline extrapolation presented here, including the *truth-tagging* uncertainties would degrade the expected limit by 4%. Including the MC statistical uncertainties would degrade the expected limit by 5%, estimated by performing a fit with scaled MC statistical uncertainties, assuming that the data-to-MC luminosity ratio stays the same as for Run 2.

Flavour tagging has a significant impact on the sensitivity of the $VH(H \to c\bar{c})$ analysis. A better flavour tagging performance can be expected [98–103] due to the improved resolution of the replacement inner detector at the HL-LHC (ITk). Using the existing c-tagging algorithm based on the DL1 algorithm and requiring a 27% c-jet efficiency operating point, initial studies suggest that the b- and light-flavour jet rejection factors are expected to improve by factors of 1.5 and 3, respectively. This performance improvement is expected to improve the $VH(H \to c\bar{c})$ limit by 10–15%, where the improvement in light-flavour rejection has a more significant impact. Additionally, the c-tagging algorithms are also expected to improve due to better machine learning approaches.

The re-interpretation of the $VH(H \to c\bar{c})$ signal strength within the κ framework allows an estimate of the sensitivity to the Higgs-charm Yukawa coupling, in terms of the coupling modifier κ_c. For the HL-LHC scenario, the resulting profile likelihood scans are shown in Fig. 4.36. The extrapolated interpretation, considering all three lepton channels, results in an expected constraint of $|\kappa_c| < 3.0$ at 95% CL.

Table 4.11 **Breakdown of contributions to the uncertainty in the fitted values of** $\mu_{VH(c\bar{c})}$ **from a fit to an Asimov dataset scaled to 3000 fb**$^{-1}$ and $\sqrt{s} = 14$ TeV. The sum in quadrature of different sources of uncertainty may differ from the total due to correlations. In the case that the up and down systematic variations have different values, the mean of the absolute values is shown

Source of uncertainty		$\Delta\mu_{VH}^{c\bar{c}}$
Total		3.21
Statistical		1.97
Systematics		2.53
Statistical uncertainties		
Data statistics only		1.59
Floating normalisations		0.95
Theoretical and modelling uncertainties		
$VH(H \rightarrow c\bar{c})$		0.27
Z+jets		1.77
Top-quark		0.96
W+jets		0.84
Diboson		0.34
$VH(H \rightarrow b\bar{b})$		0.29
Multi-Jet		0.09
Experimental uncertainties		
Jets		0.59
Leptons		0.20
$E_{\mathrm{T}}^{\mathrm{miss}}$		0.18
Pile-up and luminosity		0.19
Flavour tagging	c-jets	0.61
	b-jets	0.16
	light-jets	0.51
	τ-jets	0.19

4.11.3 Extrapolation Results of the $VH(H \rightarrow b\bar{b}/c\bar{c})$ Combination

Following the statistical combination described in Sect. 4.10, the combined fit to the $VH(H \rightarrow c\bar{c})$ and $VH(H \rightarrow b\bar{b})$ signal strength parameters can be extrapolated to the HL-LHC, resulting in expected best fit values of:

$$\mu_{VH}^{b\bar{b}} = 1.00 \pm 0.06,$$
$$\mu_{VH}^{c\bar{c}} = 1.00 \pm 3.20,$$

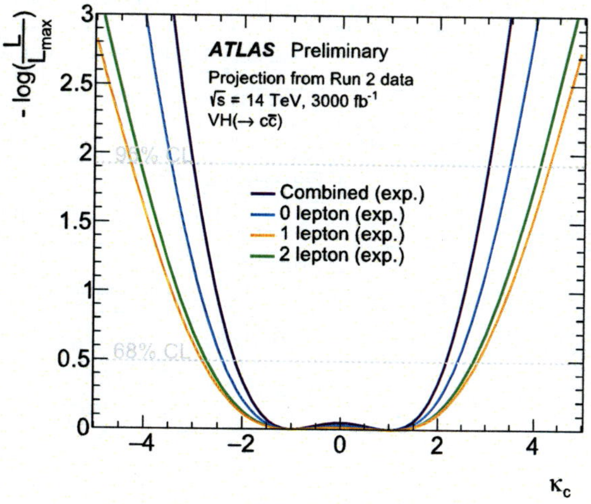

Fig. 4.36 Expected profile likelihood scans for κ_c at the HL-LHC. The per-channel likelihoods are obtained using a three POI fit, in which each channel has a separate $VH(H \to c\bar{c})$ parameter of interest

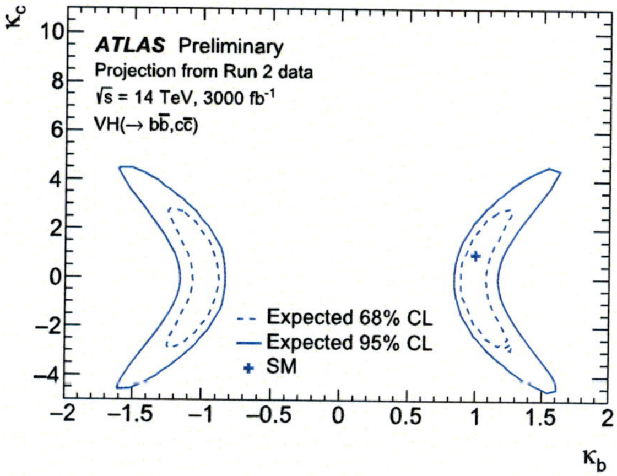

Fig. 4.37 Expected profile likelihood scans for the $VH(H \to b\bar{b}/c\bar{c})$ combination extrapolated to a dataset of 3000 fb^{-1} at $\sqrt{s} = 14$ TeV. A two-dimensional scan is shown for a simultaneous fit to κ_c and κ_b

The correlation coefficient between the $VH(H \to b\bar{b})$ and $VH(H \to c\bar{c})$ signal strength parameters is -11%. Once again, there is no significant gain compared to the individual results due to the simple nature of the combination. Reinterpreting the signal strengths in terms of κ_b and κ_c results in the likelihood scans shown in Fig. 4.37 and provides an expected constraint of $|\kappa_c/\kappa_b| < 2.7$ at 95% CL. As the value of κ_b was already constrained well from the Run 2 $VH(H \to b\bar{b})$ analysis, the improvement on the $|\kappa_c/\kappa_b|$ ratio at HL-LHC is smaller than the other quantities quoted in this study.

4.12 Conclusions

This chapter presented the recent search for $H \to c\bar{c}$ decays in the VH production mode, using the pp collision dataset collected by the ATLAS detector between 2015–2018. Events were considered in three channels, categorised by the decay of the vector boson $ZH \to \nu\nu c\bar{c}$, $WH \to \ell\nu c\bar{c}$ and $ZH \to \ell\ell c\bar{c}$. Jets originating from the decays of c-quarks are identified using machine learning algorithms known as c-tagging. Events are considered in different categories of c-tags, jet multiplicity, and p_T^V to enhance the sensitivity. The main backgrounds in the analysis are V+jets and $t\bar{t}$ events, which are constrained from data using dedicated control regions.

To validate the analysis strategy, the Standard Model diboson processes $VW(W \to cq)$ and $VZ(Z \to c\bar{c})$ were also measured. The signal strengths for these processes were found to agree with the Standard Model predictions, with observed (expected) significances of 2.6 (2.2) and 3.8 (4.6) standard deviations for the $VZ(Z \to c\bar{c})$ and the $VW(W \to cq)$ process, respectively.

The $VH(H \to c\bar{c})$ signal strength was observed to be in agreement with the SM, and an observed (expected) upper limit of 26 (31) times the predicted cross-section times branching fraction for the $VH(H \to c\bar{c})$ process was set. This result significantly improves on the previous iterations of the analysis, due to the addition of two new channels, a larger dataset, and a better c-tagger and previous results are summarised in Fig. 4.38. The result can be interpreted within the κ framework to constrain the Higgs coupling to charm quarks directly. This results in the first direct constraint on κ_c, with $|\kappa_c| < 8.5$ (12.4) observed (expected) at the 95% confidence level.

A statistical combination with the $VH(H \to b\bar{b})$ analysis was also achieved by performing a simultaneous fit to all of the $VH(H \to b\bar{b})$ and $VH(H \to c\bar{c})$ analysis categories. An upper limit on the ratio $|\kappa_c/\kappa_b|$ was found to be $|\kappa_c/\kappa_b| < 4.5$ at the

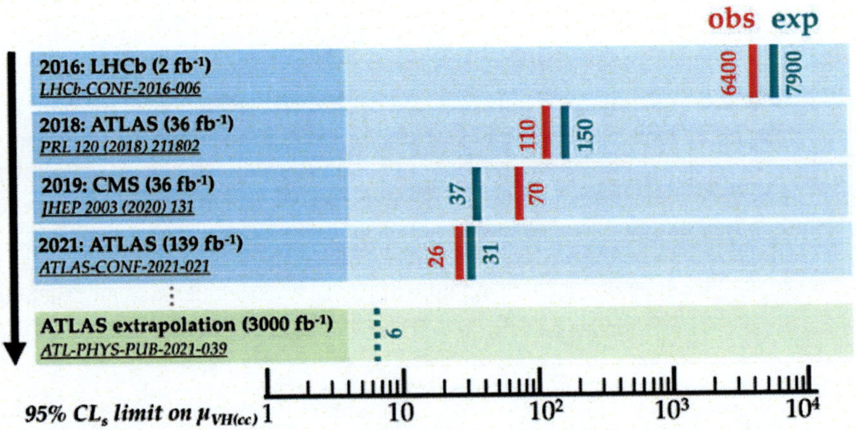

Fig. 4.38 Evolution of expected and observed upper limits on the $VH(H \to c\bar{c})$ signal strength

95% confidence level, without any assumptions on the Higgs width. This limit is smaller than the ratio of the masses of the b- and c-quarks m_b/m_c, which means that this result confirms the coupling of the Higgs boson to c-quarks to be weaker than to b-quarks. These constraints on κ_c and κ_b complement those from measurements of the Higgs boson p_T spectrum [104, 105].

The $VH(H \to c\bar{c})$ analysis was also extrapolated to the HL-LHC, following the analysis strategy of the Run 2 analysis. The expected dataset was scaled to an integrated luminosity 3000 fb^{-1} and a center-of-mass energy of 14 TeV, and the systematic uncertainties were reduced. The expected constraints in the HL-LHC scenario were found to be an upper limit on the $VH(H \to c\bar{c})$ signal strength of 6.4 times the SM prediction and a constraint of $|\kappa_c| < 3.0$. For the combination with the $VH(H \to b\bar{b})$ analysis, an upper limit (at 95% CL) of $|\kappa_c/\kappa_b| < 2.7$ is expected at the HL-LHC.

References

1. ATLAS Collaboration (2012) Observation of a new particle in the search for the Standard-Model Higgs boson with the ATLAS detector at the LHC. Phys Lett B 716:1. https://doi.org/10.1016/j.physletb.2012.08.020. arXiv: 1207.7214 [hep-ex]
2. CMS Collaboration (2012) Observation of a new boson at a mass of 125GeV with the CMS experiment at the LHC. Phys Lett B 716:30. https://doi.org/10.1016/j.physletb.2012.08.021. arXiv: 1207.7235 [hep-ex]
3. ATLAS Collaboration (2019) Cross-section measurements of the Higgs boson decaying into a pair of τ-leptons in proton. proton collisions at $\sqrt{s}$ =13TeV with the ATLAS detector. Phys Rev D 99:072001. https://doi.org/10.1103/PhysRevD.99.072001. arXiv: 1811.08856 [hep-ex]
4. ATLAS Collaboration (2018) Observation of $H \to b\bar{b}$ decays and VH production with the ATLAS detector. Phys Lett B 786:59. https://doi.org/10.1016/j.physletb.2018.09.013. arXiv: 1808.08238 [hep-ex]
5. ATLAS Collaboration (2018) Observation of Higgs boson production in association with a top quark pair at the LHC with the ATLAS detector. Phys Lett B 784:173. https://doi.org/10.1016/j.physletb.2018.07.035. arXiv: 1806.00425 [hep-ex]
6. CMS Collaboration (2018) Observation of the Higgs boson decay to a pair of τ leptons with the CMS detector. Phys Lett B 779:283. https://doi.org/10.1016/j.physletb.2018.02.004. arXiv: 1708.00373 [hep-ex]
7. CMS Collaboration (2018) Observation of $t\bar{t}H$ production. Phys Rev Lett 120:231801. https://doi.org/10.1103/PhysRevLett.120.231801. arXiv: 1804.02610 [hep-ex]
8. CMS Collaboration (2018) Observation of Higgs boson decay to bottom quarks. Phys Rev Lett 121:121801. https://doi.org/10.1103/PhysRevLett.121.121801. arXiv: 1808.08242 [hep-ex]
9. CMS Collaboration (2021) Evidence for Higgs boson decay to a pair of muons. JHEP 01:148. https://doi.org/10.1007/JHEP01(2021)148. arXiv: 2009.04363 [hep-ex]
10. ATLAS Collaboration (2021) A search for the dimuon decay of the standard model Higgs boson with the ATLAS detector. Phys Lett B 812:135980. https://doi.org/10.1016/j.physletb.2020.135980. arXiv: 2007.07830 [hep-ex]
11. ATLAS Collaboration (2018) Search for the decay of the Higgs boson to charm quarks with the ATLAS experiment. Phys Rev Lett 120:211802. https://doi.org/10.1103/PhysRevLett.120.211802. arXiv: 1802.04329 [hep-ex]

12. CMS Collaboration (2020) A search for the standard model Higgs boson decaying to charmquarks. JHEP 03:131. https://doi.org/10.1007/JHEP03(2020)131. arXiv: 1912.01662 [hep-ex]
13. ATLAS Collaboration (2020) Search for the Higgs boson decays $H \to ee$ and $H \to e\mu$ in pp collisions at $\sqrt{s}$ =13 TeV with the ATLAS detector. Phys Lett B 801:135148. https://doi.org/10.1016/j.physletb.2019.135148. arXiv: 1909.10235 [hep-ex]
14. CMS Collaboration (2015) Search for a standard model-like Higgs boson in the $\mu^+\mu^-$ and e^+e^- decay channels at the LHC. Phys Lett B 744:184. https://doi.org/10.1016/j.physletb.2015.03.048. arXiv: 1410.6679 [hep-ex]
15. ATLAS Collaboration (2018) Search for exclusive Higgs and Z boson decays to $\phi\gamma$ and $\rho\gamma$ with the ATLAS detector. JHEP 07:127. https://doi.org/10.1007/JHEP07(2018)127. arXiv: 1712.02758 [hep-ex]
16. ATLAS Collaboration (2018) Searches for exclusive Higgs and Z boson decays into $J/\psi\gamma$, $\psi(2S)\gamma$, and Υ $(nS)\gamma$ at $\sqrt{s}$ = 13TeV with the ATLAS detector. Phys Lett B 786:134. https://doi.org/10.1016/j.physletb.2018.09.024. arXiv: 1807.00802 [hep-ex]
17. CMS Collaboration (2020) Search for decays of the 125 GeV Higgs boson into a Z boson and a ρ or ϕ meson. JHEP 11:039. https://doi.org/10.1007/JHEP11(2020)039. arXiv: 2007.05122 [hep-ex]
18. CMS Collaboration (2019) Search for rare decays of Z and Higgs bosons to J/ψ and a photon in proton–proton collisions at $\sqrt{s}$ = 13TeV. Eur Phys J C 79:94. https://doi.org/10.1140/epjc/s10052-019-6562-5. arXiv: 1810.10056 [hep-ex]
19. Djouadi A, Kalinowski J, Spira M (1998) HDECAY: a program for Higgs boson decays in the standard model and its supersymmetric extension. Comput Phys Commun 108:56. https://doi.org/10.1016/S0010-4655(97)00123-9. arXiv: hep-ph/9704448
20. Delaunay C et al (2014) Enhanced Higgs boson coupling to charm pairs. Phys Rev D 89:033014. https://doi.org/10.1103/PhysRevD.89.033014
21. Perez G et al (2015) Constraining the charm Yukawa and Higgs-quark coupling universality. Phys Rev D 92:033016. https://doi.org/10.1103/PhysRevD.92.033016
22. **Botella FJ et al (2016) What if the masses of the first two quark families are not generated** by the standard model Higgs boson? Phys Rev D 94(11):115031. https://doi.org/10.1103/PhysRevD.94.115031. arXiv: 1602.08011 [hep-ph]
23. Bar-Shalom S, Soni A (2018) Universally enhanced light-quarks Yukawa couplings paradigm. Phys Rev D 98:055001. https://doi.org/10.1103/PhysRevD.98.055001
24. Ghosh D, Gupta RS, Perez G (2016) Is the Higgs mechanism of fermion mass generation a **fact? A Yukawa-less first-two-generation model. Phys Lett B 755:504–508. ISSN: 0370-2693.** https://doi.org/10.1016/j.physletb.2016.02.059
25. **Egana-Ugrinovic D, Homiller S, Meade PR (2019) Aligned and spontaneous flavor viola**tion. Phys Rev Lett 123(3):031802. https://doi.org/10.1103/PhysRevLett.123.031802. arXiv: 1811.00017 [hep-ph]
26. Egana-Ugrinovic D, Homiller S, Meade PR (2019) Higgs bosons with large couplings to light quarks. Phys Rev D 100(11):115041. https://doi.org/10.1103/PhysRevD.100.115041. arXiv: 1908.11376 [hep-ph]
27. Frixione S, Nason P, Oleari C (2007) Matching NLO QCD computations with parton shower simulations: the POWHEG method. JHEP 11:070. https://doi.org/10.1088/1126-6708/2007/11/070. arXiv: 0709.2092 [hep-ph]
28. Ball RD et al (2015) Parton distributions for the LHC run II. JHEP 04:040. https://doi.org/10.1007/JHEP04(2015)040. arXiv: 1410.8849 [hep-ph]
29. Sjöstrand T et al (2015) An introduction to PYTHIA 8.2. Comput Phys Commun 191:159. https://doi.org/10.1016/j.cpc.2015.01.024. arXiv: 1410.3012 [hep-ph]
30. ATLAS Collaboration (2014) Measurement of the Z/γ^* boson transverse momentum distribution in pp collisions at $\sqrt{s}$ = 7 TeV with the ATLAS detector. JHEP 09:145. https://doi.org/10.1007/JHEP09(2014)145. arXiv: 1406.3660 [hep-ex]
31. Ciccolini ML, Dittmaier S, Krämer M (2003) Electroweak radiative corrections to associated WH and ZH production at hadron colliders. Phys Rev D 68:073003. https://doi.org/10.1103/PhysRevD.68.073003. arXiv: hep-ph/0306234 [hep-ph]

32. Brein O, Djouadi A, Harlander R (2004) NNLO QCD corrections to the Higgs-strahlung processes at hadron colliders. Phys Lett B 579:149–156. https://doi.org/10.1016/j.physletb.2003.10.112. arXiv: hep-ph/0307206
33. Brein O et al (2012) Top-quark mediated effects in Hadronic Higgs-Strahlung. Eur Phys J C 72:1868. https://doi.org/10.1140/epjc/s10052-012-1868-6. arXiv: 1111.0761 [hep-ph]
34. Denner A et al (2015) HAWK 2.0: a Monte Carlo program for Higgs production in vectorboson fusion and Higgs strahlung at hadron colliders. Comput Phys Commun 195:161–171. https://doi.org/10.1016/j.cpc.2015.04.021. arXiv: 1412.5390 [hep-ph]
35. Brein O, Harlander RV, Zirke TJE (2013) vh@nnlo-Higgs Strahlung at hadron colliders. Comput Phys Commun 184:998–1003. https://doi.org/10.1016/j.cpc.2012.11.002. arXiv: 1210.5347 [hep-ph]
36. Cullen G et al (2012) Automated one-loop calculations with GoSAM. Eur Phys J C 72:1889. https://doi.org/10.1140/epjc/s10052-012-1889-1. arXiv: 1111.2034 [hep-ph]
37. Hamilton K, Nason P, Zanderighi G (2012) MINLO: multi-scale improved NLO. JHEP 10:155. https://doi.org/10.1007/JHEP10(2012)155. arXiv: 1206.3572 [hep-ph]
38. Luisoni G et al (2013) $HW^{\pm}/HZ + 0$ and 1 jet at NLO with the POWHEG BOX interfaced to Go Samand their merging within MiNLO. JHEP 10:083. https://doi.org/10.1007/JHEP10(2013)083. arXiv: 1306.2542 [hep-ph]
39. Altenkamp L et al (2013) Gluon-induced Higgs-strahlung at next-to-leading order QCD. JHEP 02:078. https://doi.org/10.1007/JHEP02(2013)078. arXiv: 1211.5015 [hep-ph]
40. Harlander RV et al (2014) Soft gluon resummation for gluon-induced Higgs Strahlung. JHEP 11:082. https://doi.org/10.1007/JHEP11(2014)082. arXiv: 1410.0217 [hep-ph]
41. **Frixione S, Nason P, Ridolfi G (2007) A positive-weight next-to-leading-order Monte Carlo for heavy flavour hadroproduction. JHEP 09:126.** https://doi.org/10.1088/1126-6708/2007/09/126. arXiv: 0707.3088 [hep-ph]
42. ATLAS Collaboration (2014) ATLAS Pythia 8 tunes to 7 TeV data. ATL-PHYS-PUB-2014-021. https://cds.cern.ch/record/1966419
43. Beneke M et al (2012) Hadronic top-quark pair production with NNLL threshold resummation. Nucl Phys B 855:695–741. https://doi.org/10.1016/j.nuclphysb.2011.10.021. arXiv:1109.1536 [hep-ph]
44. Cacciari M et al (2012) Top-pair production at hadron colliders with next-to-next-to-leading logarithmic soft-gluon resummation. Phys Lett B 710:612–622. https://doi.org/10.1016/j.physletb.2012.03.013. arXiv: 1111.5869 [hep-ph]
45. Bärnreuther P, Czakon M, Mitov A (2012) Percent-level-precision physics at the tevatron: next-to-next-to-leading order QCD corrections to $q\bar{q} \rightarrow t\bar{t} + X$. Phys Rev Lett 109:132001. https://doi.org/10.1103/PhysRevLett.109.132001. arXiv: 1204.5201 [hep-ph]
46. Czakon M, Mitov A (2012) NNLO corrections to top-pair production at hadron colliders: the all-fermionic scattering channels. JHEP 12:054. https://doi.org/10.1007/JHEP12(2012)054. arXiv: 1207.0236 [hep-ph]
47. Czakon M, Mitov A (2013) NNLO corrections to top pair production at hadron colliders: the quark-gluon reaction. JHEP 01:080. https://doi.org/10.1007/JHEP01(2013)080. arXiv: 1210.6832 [hep-ph]
48. Czakon M, Fiedler P, Mitov A (2013) Total top-quark pair-production cross section at Hadron colliders through $O(\alpha_S^4)$. Phys Rev Lett 110:252004. https://doi.org/10.1103/PhysRevLett.110.252004. arXiv: 1303.6254 [hep-ph]
49. Czakon M, Mitov A (2014) Top++: a program for the calculation of the top-pair cross-section at hadron colliders. Comput Phys Commun 185:2930. https://doi.org/10.1016/j.cpc.2014.06.021. arXiv: 1112.5675 [hep-ph]
50. Alioli S et al (2009) NLO single-top production matched with shower in POWHEG: s - and t - channel contributions. JHEP 09:111. https://doi.org/10.1088/1126-6708/2009/09/111. arXiv: 0907.4076 [hep-ph]. Erratum: Alioli et al (2010) JHEP 02:011. https://doi.org/10.1007/JHEP02(2010)011
51. Aliev M et al (2011) HATHOR – HAdronic top and heavy quarks crOss section calculatoR. Comput Phys Commun 182:1034–1046. https://doi.org/10.1016/j.cpc.2010.12.040. arXiv: 1007.1327 [hep-ph]

52. Kant P et al (2015) HatHor for single top-quark production: Updated predictions and uncertainty estimates for single top-quark production in hadronic collisions. Comput Phys Commun 191:74–89. https://doi.org/10.1016/j.cpc.2015.02.001. arXiv: 1406.4403 [hep-ph]
53. Re E (2011) Single-top Wt-channel production matched with parton showers using the POWHEG method. Eur Phys J C 71:1547. https://doi.org/10.1140/epjc/s10052-011-1547-z. arXiv: 1009.2450 [hep-ph]
54. Kidonakis N (2010) Two-loop soft anomalous dimensions for single top quark associated production with a W^- or H^-. Phys Rev D 82:054018. https://doi.org/10.1103/PhysRevD.82.054018. arXiv: 1005.4451 [hep-ph]
55. Kidonakis N (2013) Top quark production. In: Proceedings, helmholtz international summer school on physics of heavy quarks and hadrons (HQ 2013) (JINR, Dubna, Russia, July 15–28, 2013), pp 139–168. https://doi.org/10.3204/DESY-PROC-2013-03/Kidonakis. arXiv: 1311.0283 [hep-ph]
56. Gleisberg T et al (2009) Event generation with SHERPA 1.1. JHEP 02:007. https://doi.org/10.1088/1126-6708/2009/02/007. arXiv: 0811.4622 [hep-ph]
57. Höche S et al (2009) QCD matrix elements and truncated showers. JHEP 05:053. https://doi.org/10.1088/1126-6708/2009/05/053. arXiv: 0903.1219 [hep-ph]
58. Cascioli F, Maierhöfer P, Pozzorini S (2012) Scattering amplitudes with open loops. Phys Rev Lett 108:111601. https://doi.org/10.1103/PhysRevLett.108.111601. arXiv: 1111.5206 [hep-ph]
59. Anastasiou C et al (2004) High precision QCD at hadron colliders: electroweak gauge boson rapidity distributions at next-to-next-to leading order. Phys Rev D 69:094008. https://doi.org/10.1103/PhysRevD.69.094008. arXiv: hep-ph/0312266
60. Alioli S et al (2010) A general framework for implementing NLO calculations in shower Monte Carlo programs: the POWHEG BOX. JHEP 06:043. https://doi.org/10.1007/JHEP06(2010)043. arXiv: 1002.2581 [hep-ph]
61. LHC Higgs Cross SectionWorking Group et al (2013) Handbook of LHC Higgs cross sections: 3. Higgs properties. CERN-2013-004. https://doi.org/10.5170/CERN-2013-004. arXiv:1307.1347 [hep-ph]
62. Alwall J et al (2014) The automated computation of tree-level and next-to-leading order differential cross sections, and their matching to parton shower simulations. JHEP 07:079. https://doi.org/10.1007/JHEP07(2014)079. arXiv: 1405.0301 [hep-ph]
63. Bellm J et al (2016) Herwig 7.0/Herwig++ 3.0 release note. Eur Phys J C 76(4):196. https://doi.org/10.1140/epjc/s10052-016-4018-8. arXiv: 1512.01178 [hep-ph]
64. ATLAS Collaboration (2015) Vertex reconstruction performance of the ATLAS detector at $\sqrt{s} = 13TeV$. ATL-PHYS-PUB-2015-026. https://cds.cern.ch/record/2037717
65. ATLAS Collaboration (2019) Electron and photon performance measurements with the ATLAS detector using the 2015–2017 LHC proton-proton collision data. JINST 14:P12006. https://doi.org/10.1088/1748-0221/14/12/P12006. arXiv: 1908.00005 [hep-ex]
66. **ATLAS Collaboration (2021) Muon reconstruction and identification efficiency in ATLAS** using the full Run 2 pp collision data set at $\sqrt{s}$ =13TeV. Eur Phys J C 81:578. https://doi.org/10.1140/epjc/s10052-021-09233-2. arXiv: 2012.00578 [hep-ex]
67. ATLAS Collaboration (2016) Reconstruction of hadronic decay products of tau leptons with the ATLAS experiment. Eur Phys J C 76:295. https://doi.org/10.1140/epjc/s10052-016-4110-0. arXiv: 1512.05955 [hep-ex]
68. **ATLAS Collaboration (2017) Measurement of the tau lepton reconstruction and identification** performance in the ATLAS experiment using pp collisions at $\sqrt{s} = 13TeV$. ATLAS-CONF-2017- 029. https://cds.cern.ch/record/2261772
69. ATLAS Collaboration (2018) Performance of missing transversemomentumreconstruction with the ATLAS detector using proton-proton collisions at $\sqrt{s}$ = 13TeV. Eur Phys J C 78:903. https://doi.org/10.1140/epjc/s10052-018-6288-9. arXiv: 1802.08168 [hep-ex]
70. ATLAS Collaboration (2017) Performance of the ATLAS track reconstruction algorithms in dense environments in LHC Run 2. Eur Phys J C 77:673. https://doi.org/10.1140/epjc/s10052-017-5225-7. arXiv: 1704.07983 [hep-ex]

71. ATLAS Collaboration (2017) Jet energy scale measurements and their systematic uncertainties in proton-proton collisions at $\sqrt{s}$ = 13TeV with the ATLAS detector. Phys Rev D 96:072002. https://doi.org/10.1103/PhysRevD.96.072002. arXiv: 1703.09665 [hep-ex]
72. ATLAS Collaboration (2017) Topological cell clustering in the ATLAS calorimeters and its performance in LHC Run 1. Eur Phys J C 77:490. https://doi.org/10.1140/epjc/s10052-017-5004-5. arXiv: 1603.02934 [hep-ex]
73. ATLAS Collaboration (2015) Properties of jets and inputs to jet reconstruction and calibration with the ATLAS detector using proton-proton collisions at $\sqrt{s} = 13TeV$. ATL-PHYS-PUB-2015-036. https://cds.cern.ch/record/2044564
74. Cacciari M, Salam GP, Soyez G (2008) The anti-k_t jet clustering algorithm. JHEP 04:063. https://doi.org/10.1088/1126-6708/2008/04/063. arXiv: 0802.1189 [hep-ph]
75. Cacciari M, Salam GP, Soyez G (2012) FastJet user manual. Eur Phys J C 72:1896. https://doi.org/10.1140/epjc/s10052-012-1896-2. arXiv: 1111.6097 [hep-ph]
76. ATLAS Collaboration (2016) Performance of pile-up mitigation techniques for jets in pp collisions at $\sqrt{s}$ = 8TeV using the ATLAS detector. Eur Phys J C 76:581. https://doi.org/10.1140/epjc/s10052-016-4395-z. arXiv: 1510.03823 [hep-ex]
77. ATLAS Collaboration (2021) Measurements of WH and ZH production in the $H \rightarrow b\bar{b}$ decay channel in pp collisions at 13TeV with the ATLAS detector. Eur Phys J C 81:178. https://doi.org/10.1140/epjc/s10052-020-08677-2. arXiv: 2007.02873 [hep-ex]
78. **ATLAS Collaboration (2019) ATLAS b-jet identification performance and efficiency measurement** with $t\bar{t}$ events in pp collisions at $\sqrt{s}$ = 13TeV. Eur Phys J C 79:970. https://doi.org/10.1140/epjc/s10052-019-7450-8. arXiv: 1907.05120 [hep-ex]
79. **ATLAS Collaboration (2020) Optimisation of the smoothing of b -jet identification efficiency** and mistag rate simulation-to-data scale factors in ATLAS. Technical report, CERN, Geneva. https://cds.cern.ch/record/2710598
80. Bell AS (2016) Calibration of b -tagging using di-leptonic $t\bar{t}$ events produced in pp collisions at $\sqrt{s}$ = 13 TeV and a combinatorial likelihood approach. Technical report ATL-COM-PHYS-2016-1598. CERN, Geneva. https://cds.cern.ch/record/2231568
81. **Guindon S et al (2017) Measurement of c -jet tagging efficiency in** $t\bar{t}$ events using a likelihood approach. Technical report ATL-COM-PHYS-2017-073. CERN, Geneva. https://cds.cern.ch/record/2243764
82. Erdmann J et al (2014) A likelihood-based reconstruction algorithm for top-quark pairs and the KLFitter framework. Nucl Instrum Meth A 748(18):25. ISSN: 0168-9002. https://doi.org/10.1016/j.nima.2014.02.029
83. **ATLAS Collaboration (2018) Calibration of light-flavour b-jet mistagging rates using ATLAS** proton–proton collision data at $\sqrt{s} = 13TeV$. ATLAS-CONF-2018-006. https://cds.cern.ch/record/2314418
84. ATLAS Collaboration (2019) Luminosity determination in pp collisions at $\sqrt{s} = 13TeV$ using the ATLAS detector at the LHC
85. ATLAS Collaboration (2018) Search for the standard model Higgs boson produced in association with a vector boson and decaying to a pair of b-quarks. Technical report ATL-COM-PHYS-2018-512. CERN, Geneva. https://cds.cern.ch/record/2317111
86. **Collaboration ATLAS (2019) Electron reconstruction and identification in the ATLAS experiment** using the 2015 and 2016 LHC proton-proton collision data at $\sqrt{s}$ =13 TeV. Eur Phys J C 79(8):639
87. Cranmer K et al (2012) HistFactory: a tool for creating statistical models for use with RooFit and RooStats
88. Verkerke W, Kirkby D (2003) The RooFit toolkit for data modeling. arXiv: physics/0306116 [physics.data-an]
89. Moneta L et al (2010) The RooStats project. PoS ACAT2010:057. https://doi.org/10.22323/1.093.0057. arXiv: 1009.1003 [physics.data-an]
90. Read AL (2002) Presentation of search results: the CL_S technique. J Phys G 28:2693. https://doi.org/10.1088/0954-3899/28/10/313

91. Cowan G et al (2011) Asymptotic formulae for likelihood-based tests of new physics. Eur Phys J C 71:1554. https://doi.org/10.1140/epjc/s10052-011-1554-0. arXiv: 1007.1727 [physics.data-an]. Erratum: Cowan G et al (2013) Eur Phys J C 73, 2501. https://doi.org/10.1140/epjc/s10052-013-2501-z
92. de Florian D et al (2016) Handbook of LHC Higgs cross sections: 4. Deciphering the Nature of the Higgs sector. CERN Yellow reports: monographs. CERN, Geneva. https://doi.org/10.23731/CYRM-2017-002
93. Badger S et al (2016) Les houches 2015: physics at TeV colliders standard model working group report. https://doi.org/10.48550/ARXIV.1605.04692
94. Berger N et al (2019) Simplified template cross sections - stage 1.1. https://doi.org/10.48550/ARXIV.1906.02754. https://arxiv.org/abs/1906.02754
95. ATLAS Collaboration (2020) A combination of measurements of Higgs boson production and decay using up to $139 fb^{-1}$ of proton–proton collision data at $\sqrt{s} = 13TeV$ collected with the ATLAS experiment. ATLAS-CONF-2020-027. https://cds.cern.ch/record/2725733
96. Bazavov A et al (2018) Up-, down-, strange-, charm-, and bottom-quark masses from four-flavor lattice QCD. Phys Rev D 98:054517. https://doi.org/10.1103/PhysRevD.98.054517
97. ATLAS Collaboration (2021) Extrapolation of ATLAS sensitivity to $H \rightarrow b\bar{b}$ and $H \rightarrow c\bar{c}$ decays in V H production at the HL-LHC. Technical report CERN, Geneva. http://cds.cern.ch/record/2788490
98. ATLAS Collaboration (2021) Expected performance of the ATLAS detector under different high-luminosity LHC conditions. ATL-PHYS-PUB-2021-023. http://cds.cern.ch/record/2776650
99. ATLAS Collaboration (2019) Expected performance of the ATLAS detector at the high-luminosity LHC. ATL-PHYS-PUB-2019-005. https://cds.cern.ch/record/2655304
100. ATLAS Collaboration (2017) Technical design report for the ATLAS inner tracker pixel detector. https://cds.cern.ch/record/2285585
101. ATLAS Collaboration (2017) Technical design report for the ATLAS inner tracker strip detector. http://cds.cern.ch/record/2257755
102. ATLAS Collaboration (2020) Expected b-tagging performance with the upgraded ATLAS inner tracker detector at the high-luminosity LHC. ATL-PHYS-PUB-2020-005. https://cds.cern.ch/record/2713377
103. Expected tracking and related performance with the updated ATLAS inner tracker layout at the high-luminosity LHC. Technical report CERN, Geneva, July 2021. http://cds.cern.ch/record/2776651
104. ATLAS Collaboration (2020) Measurements of the Higgs boson inclusive and differential fiducial cross sections in the $4l$ decay channel at $\sqrt{s} = 13$TeV. Eur Phys J C 80:942. https://doi.org/10.1140/epjc/s10052-020-8223-0. arXiv: 2004.03969 [hep-ex]
105. ATLAS Collaboration (2021) Measurements of the Higgs boson inclusive and differential fiducial cross-sections in the diphoton decay channel with $139 fb^{-1}$ of pp collisions at $\sqrt{s} = 13TeV$ with the ATLAS detector

Chapter 5
Silicon Pixel Detectors

In this chapter, a brief introduction to silicon detectors is given. The aim is to introduce the necessary background for the measurements presented in Chaps. 6 and 7. Further information ion silicon detectors can be found in [1–3].

5.1 Interactions of Particles with Matter

5.1.1 Massive Particles

Particles are detected through their interactions with matter. For example, in semiconducting detectors, charged particles interact with the material through the Coulomb interaction. This interaction ionises the atoms, moving the electrons to the conduction band and leaving holes in the valence band [3]. For a charged particle, the energy loss per unit distance due to ionisation can be described by the Bethe-Bloch formula [4]:

$$-\frac{1}{\rho}\left\langle\frac{dE}{dx}\right\rangle = 2\pi N_A r_e^2 m_e c^2 \frac{Z}{A}\frac{z^2}{\beta^2}\left\{\ln\left(\frac{2m_e c^2\beta^2\gamma^2 W_{max}}{I^2}\right) - 2\beta^2 - \delta - 2\frac{C}{Z}\right\} \tag{5.1}$$

where N_A is Avogadro's constant, r_e is the classical electron radius, m_e the mass of the electron, Z the atomic number of the absorbing material, A the atomic mass number, z is the charge of the incoming particle in units of e, I the mean excitation potential, W_{max} the maximum energy transfer in a single collision, C the shell correction, δ the density correction and ρ is the density of the material. The first two terms describe the full quantum mechanical derivation of energy loss due to ionisation. The third term δ describes density corrections to the energy loss due to polarisation, which are relevant at high energies. The final term C/Z describes shell corrections

M. Mironova, *Search for Higgs Boson Decays to Charm Quarks with the ATLAS Experiment and Development of Novel Silicon Pixel Detectors*, Springer Theses,
https://doi.org/10.1007/978-3-031-36220-0_5

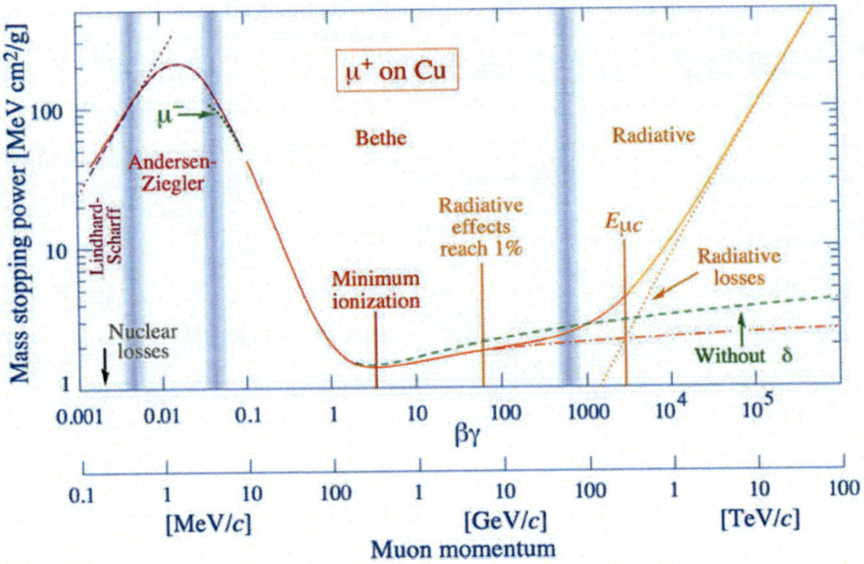

Fig. 5.1 Average energy loss approximated by the Bethe-Bloch formula for a muon across a copper medium. From [4]

due to atomic binding energies that have been neglected and which are particularly important at low energies. The approximate energy loss of a muon travelling through copper is shown in Fig. 5.1.

The Bethe-Bloch curve has a minimum at $\beta\gamma = 3.5$. Particles with such a $\beta\gamma$ are called minimum ionising particles (MIPs), as they only lose a minimum amount of kinetic energy while they traverse the material. The stopping power for a MIP in silicon is [4]:

$$\left\langle \frac{dE}{dx} \right\rangle = 3.88\ \frac{\text{MeV}}{\text{cm}}. \tag{5.2}$$

The actual energy deposited by MIPs is probabilistic in nature and can be described by Landau distributions, examples of which are shown in Fig. 5.2 for different thicknesses of silicon. The peak of the distribution is called the Most Probable Value (MPV). A characteristic feature of the Landau distribution is the large tail of higher energy depositions caused by δ-electrons. Particles with $1 < \beta\gamma < 10$ are usually referred to as MIPs. Hadrons and muons in the LHC typically fall within this range, so detector response is often characterised for MIPs. Electron energy deposition is more complicated, as their $\beta\gamma$ is much larger, and their energy loss is dominated by other processes such as Bremsstrahlung.

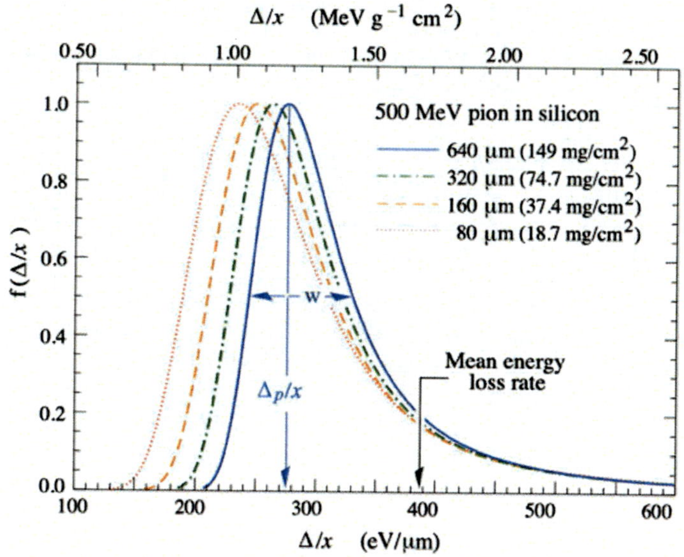

Fig. 5.2 Landau distributions from a 500 MeV pion interacting with different thicknesses of silicon. The distributions are scaled so that the most probable value is 1. From [4]

5.1.2 Photons

Photons do not have mass, and interact with materials through three main processes: the photoelectric effect, Compton scattering or pair production [6]. These effects dominate depending on the material and, more importantly, on the photon energy. The three processes are shown in Fig. 5.3.

Photoelectric effect The atom absorbs the incident photon, and the energy of the photon, E_γ is transferred onto an electron in one of the inner shells of the atoms, which has a binding energy E_{bind}. The electron is then emitted from the atom with kinetic energy $T = E_\gamma - E_{\mathrm{bind}}$. This electron can then produce further ionisation along its track, e.g. electron-hole pairs in the case of a semiconductor. In addition to

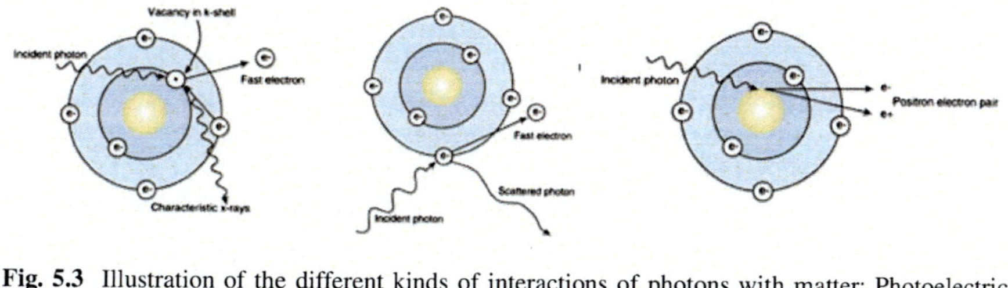

Fig. 5.3 Illustration of the different kinds of interactions of photons with matter: Photoelectric effect (left), Compton scattering (center) and Pair production (right). From [5]

this, an electron from an outer shell can fill the vacancy in the inner shell of the atom while emitting another photon. The photoelectric effect is the dominant process up to 50 keV, depending on the material.

Compton scattering In Compton scattering, the photon scatters inelastically from one of the electrons in the outer shell of the atom. This electron has small binding energy and is ejected from the atom with the additional energy from the photon, while the photon is scattered, keeping its remaining energy. This process is dominant between roughly 50 keV and 1.5 MeV.

Pair-production A highly energetic photon is converted into an electron-positron pair in the Coulomb field of a nucleus. This is only possible once the photon energy exceeds twice the electron rest mass, i.e. $E_\gamma > 1.02$ MeV, and the process becomes dominant above roughly 1.5 MeV.

When a photon beam enters a material, the photons interact with a material with a certain probability. After the interaction, the intensity of the beam decreases according to the Beer-Lambert law [8]:

$$I(x) = I_0 e^{-\mu x} \tag{5.3}$$

where x is the distance travelled, I_0 is the initial intensity, and μ is the material-specific and energy-dependent mass attenuation coefficient. The attenuation coefficient takes into account all three interaction processes described above, i.e. $\mu = \mu_{\text{photo}} + \mu_{\text{compton}} + \mu_{\text{pair}}$ [7]. The attenuation probability due to the photoelectric effect decreases with photon energy and is strongly dependent on the atomic number Z. In addition to this, the attenuation coefficient shows jumps at particular material-dependent energies, known as absorption edges. The absorption edges are caused by the photoelectric effect, where the energy of the photon exceeds the binding energy of an atomic shell. An example of the absorption coefficient in Silicon is shown in Fig. 5.4.

5.2 Semiconductors and pn-Junctions

In a lattice, the orbits of closely bound atoms can overlap, and the energy levels of a single atom can shift to be separated into two energy bands that the electrons can occupy. These are referred to as the *valence* and *conduction band*. The valence electrons are bound to the atoms, whereas electrons in the conduction band can move freely across the material. The energy gap between the conduction and valence band is known as the *band gap*. The occupancy of each band can be described by the Fermi-Dirac distribution as:

$$f(E, T) = \frac{1}{e^{(E-E_f)/k_B T} + 1} \tag{5.4}$$

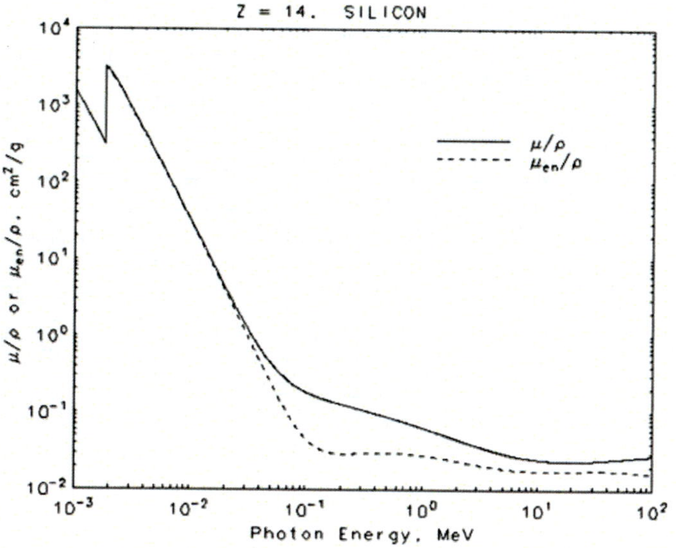

Fig. 5.4 The mass attenuation coefficient (μ/ρ) and the mass attenuation energy-absorption coefficient (μ_{en}/ρ) as a function of energy of the incident photon for silicon. From [7]

where T is the temperature, k_B is the Boltzmann constant, and E_f is the Fermi level, which is where the occupancy would be half, usually halfway between the valence and conduction band.

Materials can be characterised by the size of the band gap. For a conductor, the valence and conduction band overlap, allowing electrons to freely move between atoms, which explains the good conductivity of metals. For insulators, the valence and conduction bands are separated by a large bandgap of more than 3 eV. For a semiconductor, the bandgap is usually less than 3 eV, meaning that electrons from the valence band can be easily promoted to the conduction band through excitation. When an electron moves to the conduction band, it leaves a vacancy for one electron in the valence band, which is known as a *hole*. The hole can also move freely and transfer a positive charge and can be filled by another electron in a process known as *recombination*.

Doping of Silicon The semiconductor properties of silicon can be modified by replacing silicon atoms (*doping*) with different elements (*dopant atoms*), providing additional electrons or holes in the lattice. The effect of doping is a shift in the Fermi level of the material towards the valence or conduction band. Silicon has 4 valence electrons, so the doping is usually done using elements with 3 or 5 valence electrons. The effect of doping is illustrated in Fig. 5.5. Phosphorous is commonly used as a dopant. It has five valence electrons and can act as a donor, i.e. provide an additional free electron. This results in so-called *n-type* silicon, where electrons are the majority charge carrier. Boron is also a commonly used dopant with 3 valence electrons, which

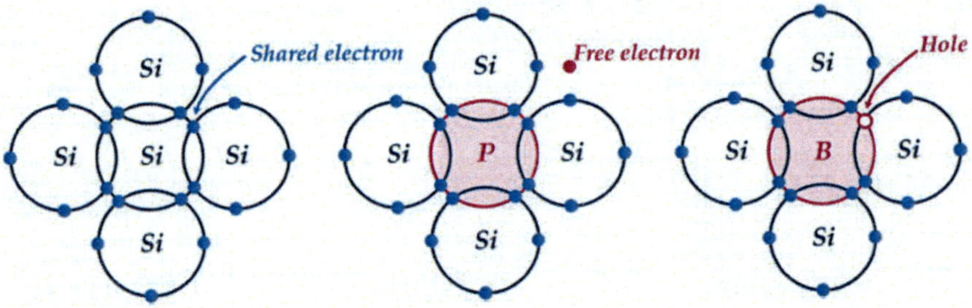

Fig. 5.5 Electronic structure of pure silicon (left), n-doped (center) and p-doped (right) silicon lattices

means it can act as an acceptor by providing an additional vacancy in the lattice. This results in *p-type* silicon, where holes are the majority charge carrier. The majority charge carriers can diffuse freely through the silicon lattice at room temperature.

pn-junctions The *n-type* and *p-type* silicon can be placed together to form a *pn-junction*. In this configuration, the majority charge carriers become attracted to their oppositely charged counterparts. The electrons and holes start to diffuse and recombine in the p-type or n-type silicon. This leads to charge building up, negative charge in the p-type silicon and positive charge in the n-type silicon, as illustrated in Fig. 5.6. This results in an electric field, which opposes the diffusion direction. After some time, an equilibrium between the E field and the diffusion of charge carriers is reached resulting in a region in the silicon with no free charge carriers and an intrinsic electric field. This is known as the *depletion region*. The electric field E and potential ϕ in the depletion region can be calculated from the charge density ρ by solving Poisson's equation $\nabla^2\phi = -\frac{\rho}{\epsilon}$. The results are illustrated in the bottom panels of Fig. 5.6. It is possible to calculate the intrinsic electric field across the depletion region, which is known as the built-in voltage ($\Delta V_{\text{internal}}$) [1]:

$$\Delta V_{\text{internal}} = \frac{kT}{e} \ln\left(\frac{N_d N_a}{n_d n_a}\right), \tag{5.5}$$

where N_d and N_a are densities of the donors and acceptors, and n_d and n_a are the numbers of free carriers of each type. It is also possible to calculate the width of the depletion region by solving Poisson's equation for the potential while taking into account the charge carrier density. This results in an expression for the depletion depth D of

$$D = \sqrt{\Delta V \frac{2\epsilon_0 \epsilon_r}{e}\left(\frac{1}{N_d} + \frac{1}{N_a}\right)} \approx \sqrt{\Delta V \frac{2\epsilon_0 \epsilon_r}{e N_{eff}}}, \tag{5.6}$$

where again N_d and N_a are densities of the donors and acceptors and ϵ_r is the permittivity of the material. In many practical cases, one type of dopant will strongly outnumber the other, in which case the effective dopant density N_{eff} can be used, which

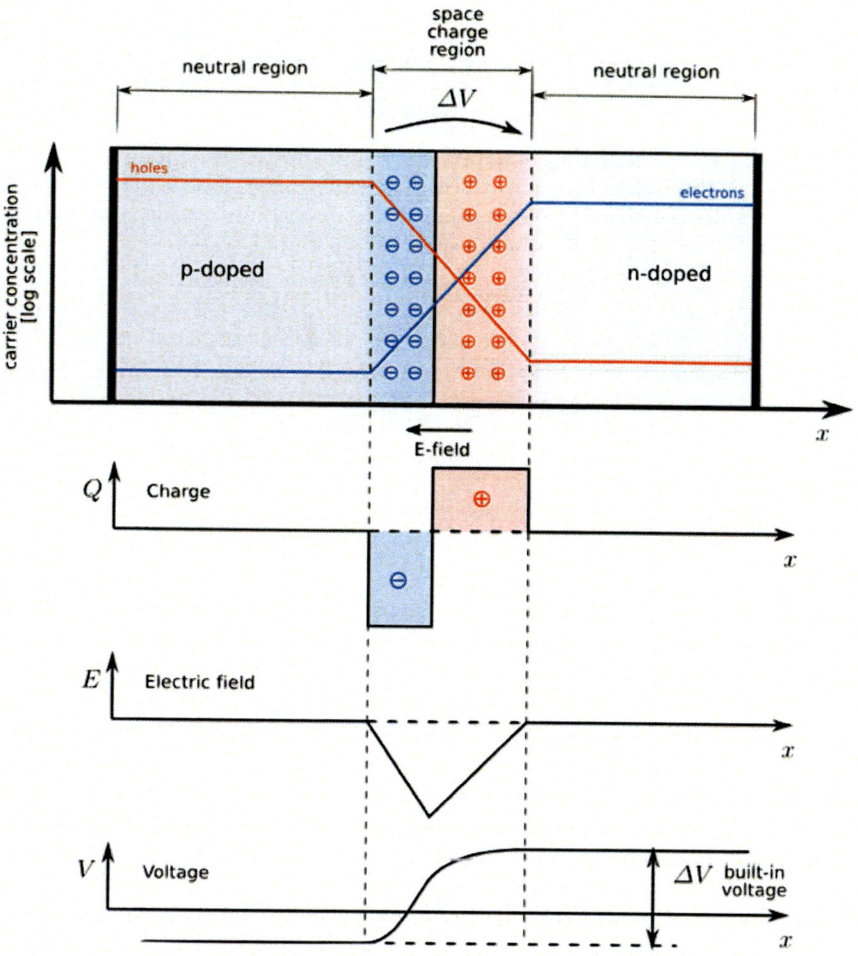

Fig. 5.6 Illustration of p-type and n-type silicon forming a pn-junction. The distribution of charge, electric field and voltage are shown in the bottom panels. From [9]

corresponds to $N_a - N_d$. Furthermore, the resistivity of material can be defined as $\rho = 1/e\mu|N_{eff}|$, where μ is the mobility of the charge carriers. Using the resistivity, the equation for the depletion depth simplifies to

$$D = \sqrt{\Delta V 2\epsilon_0\epsilon_r\mu\rho}. \tag{5.7}$$

Typically, the intrinsic voltage is of the order of 1 V and the depletion depth roughly a few micrometres, which is significantly smaller than the thickness of silicon used for detectors (~100–300 μm). The size of the depletion depth can be modified by applying an external voltage. *Forward biasing* corresponds to applying a voltage where the positive node is applied to the p-type region, and the negative node is applied to the n-type region. This counteracts the intrinsic electric field and reduces the size of the depletion region. *Reverse biasing* is the opposite, and increases the depletion depth as $D = \sqrt{(V_{\text{external}} + \Delta V)2\epsilon_0\epsilon_r\mu\rho}$. Reverse biased silicon forms the basis for particle detection using silicon. One of the common characteristics of a pn-junction is its current as a function of voltage, also known as an IV curve. In the forward bias mode, the junction will become conductive very quickly. In reverse biasing mode, there is a small current through the junction, which is known as the *leakage current* and is mostly caused by thermal excitation. At high voltages, the pn-junction can also become conductive in reverse bias mode, in a mechanism known as *breakdown*. The breakdown is caused by different mechanisms such as thermal instability (a local rise in temperature), avalanche multiplication and Zener breakdown. For the latter, the field within the junction is high enough to increase the probability that charges tunnel from the valence to the conduction band.

Signal generation A charged particle traversing a semiconductor can excite electrons and produce electron-hole pairs. For a semiconductor, the amount of signal charge carriers produced by a MIP is Q with:

$$Q = \frac{\frac{dE}{dx}e\rho d}{I_0} \tag{5.8}$$

where I_0 is the energy loss required to lift an electron into the conduction band and create an electron-hole pair, d is the thickness of the detector and ρ is the density of the material. For silicon, $I_0 = 3.63$ eV and $\rho = 2.329$ cm^{-3}g, which leads to $Q = 24000e$ for $d = 250$ μm [1, 10].

The created charge carriers are then free to move and are affected by the internal or external electric field, moving along the field lines. The movement of charge carriers is known as *drift*. In a constant electric field (<100 kV/cm), the charge carriers can be assumed to have constant mobility μ, which depends on the material properties and the type of charge carrier [11]. From the mobility of electrons (μ_e) and holes (μ_h), the drift velocity can be calculated as a function of the applied electric field $v_{\text{Drift}} = \mu_{e/h}E$.

When a charge moves towards an electrode, it induces a current in the electrode, which can be measured. The current starts instantaneously as the charges begin separating and stops when the charges reach the electrodes or when they recombine. The instantaneous induced current i_k on the electrode can be calculated according to the Shockley-Ramo theorem [12]:

$$i_k = q\vec{v}\vec{E_W} \tag{5.9}$$

with $\vec{v}$ being the drift velocity of the charge carriers, q the total deposited charge and E_W the weighting field, which can be derived from the voltage at the electrodes and the geometrical configuration of the detector. The weighting field is a virtual field calculated when the charge is removed, a voltage of 1 V is applied to the collection electrode, and 0 V is applied to all the other electrodes.

5.3 Pixel Detectors

As described above, pn-junctions can be used to detect charged particles. Segmenting the pn-junction in either one or two dimensions allows measuring the position of a particle, for example, in the ATLAS detector [13]. Devices with segments in one direction are known as *strip detectors*. Those with two-dimensional segmentation are *pixel detectors*, which are the focus of this thesis. Pixel detectors can be constructed in several ways. They are commonly built as hybrid pixel detectors, with a schematic shown in Fig. 5.8. This means that the active area and the readout electronics are located in different pieces of silicon. The active area, where the depletion region is used to collect charge, is the *sensor*. The processing of the electronic signals is done in the *front-end chip* (or ASIC), and the two components are connected pixel-by-pixel using *bump bonds*. The readout chip processes the current induced by the charge carriers further, turning it into a digital signal, with the details for the ATLAS readout chip described in Sect. 6.3 [14]. The digital signal is then usually transferred to a printed circuit from the readout chip, also known as the *flex*, which is used to further transmit the information out of the detector. The flex and the readout chip are connected using wirebonds (Fig. 5.7).

It is also possible to build pixel detectors as monolithic detectors, where the active area and readout electronics are located in the same piece of silicon. This technology is not as advanced and radiation tolerant as hybrid pixel detectors but can have other advantages. Monolithic pixel detectors are discussed further in Chap. 7.

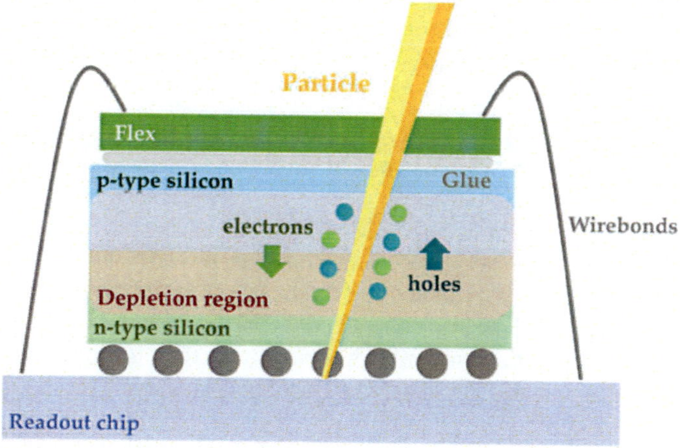

Fig. 5.7 Illustration of a hybrid pixel detector (not to scale). The signal generation in the depletion region is shown, as well as the attached readout chip and flex PCB

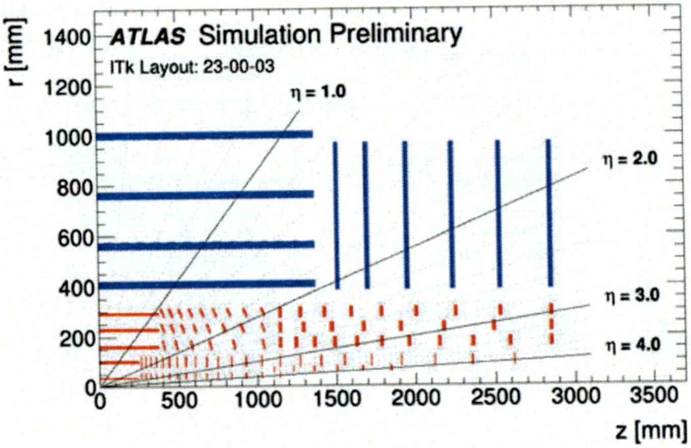

Fig. 5.8 A schematic depiction of the ITk Layout. The strip detector is shown in blue and the pixel detector is shown in red. From [23]

5.4 Radiation Damage

Silicon detector prototypes are often irradiated with protons or neutrons to simulate and study the expected behaviour of the detector after operation in a high-radiation environment, such as in the ATLAS Inner Tracker. The dose is usually delivered over a short period, and the detectors are subsequently tested. Radiation damage affects all areas of detector performance, so a good understanding of detector properties after irradiation is vital. There are two types of radiation damage: *TID damage* and *bulk damage*.

Bulk damage Bulk damage is caused by non-ionising energy loss (NIEL), in which a particle traverses the material and causes different kinds of defects [15]. For example, the particle can have enough energy to displace a silicon atom from its place in the lattice, leaving an empty lattice location (*vacancy*) and a displaced atom (*interstitial*). The interstitials or vacancies can also form interactions with impurity atoms, leading to additional defects with different properties. NIEL damage is commonly given in units of 1 MeV neutron equivalent per area (n_{eq}/cm^2). Bulk damage can cause a change in the effective space charge density and, therefore, the doping concentration, as well as an increase in leakage current and a decrease in charge collection efficiency. The defects introduced in the bulk can trap charge carriers as they travel toward the collection electrode. While neutrons mainly cause NIEL damage, protons can cause a combination of TID and bulk damage at rates depending on their energy.

TID damage Total ionising dose (TID) predominantly affects the front-end electronics in the readout chips and sensor surface properties. The surface of a readout chip usually contains metal layers for data transfer and a silicon dioxide (SiO_2) dielectric layer between the metal layers and the silicon [13]. TID damage mainly affects the dielectric and the boundary between the dielectric layer and the silicon. The atoms in these regions are not ordered regularly in a lattice, so bulk damage effects are insignificant. However, ionising radiation can create electron-hole pairs in the silicon dioxide. A large fraction of the created charge carriers recombine, but some will remain and travel in the material. Electrons are usually collected quickly due to their high mobility. However, holes travel slower and have a higher probability of getting stuck on the boundary between the silicon and the oxide, increasing the total charge in the dielectric. Additionally, clustered charges near the surface can also impact the readout chip performance. Overall, TID damage can modify the behaviour of the transistors used in the implementation of the readout logic and increase the noise in the detector. TID damage is usually given in units of Rad.

In transistors using 65 nm CMOS technology, e.g. in the ATLAS ITk Pixel readout chip, TID damage can occur in different areas, and due to the small size of the transistors, edge effects make up the most significant contribution [16]. For instance, charge build-up in the Shallow Trench Isolation (STI) and spacers can significantly change the leakage current and threshold voltage of the transistor [17, 18]. Irradiation studies of single transistors show that the radiation tolerance is worse for narrow-channel transistors, and more damage is observed at high temperatures and at low dose rates [19, 20].

After irradiation, the silicon properties do not stay constant but change over time. This process is known as *annealing* [1]. A vacancy can recombine with an interstitial or the defect can transform into a different kind of defect with new properties. Over a short period of time, on the timescale of weeks at room temperature, the radiation damage effects have been shown to decrease due to annealing (*beneficial annealing*). After more time, however, the opposite is observed: the space charge starts to increase, which has a detrimental effect on detector performance (*reverse annealing*). Annealing is expected to happen more quickly at higher temperatures [15]. Irradiated samples are usually kept at low temperatures to avoid changing detector properties through annealing.

5.5 The ITk Pixel Upgrade

The LHC started operating in 2008 and has had a hugely successful run, with a total dataset of almost 200 fb^{-1} delivered to both the ATLAS and CMS experiments during Run 1 and Run 2. In the coming years, during Run 3 of the LHC, an even larger dataset will be delivered. After the end of Run 3 in 2026, the LHC will undergo a significant upgrade to exploit its physics potential fully. The upgrade, which is scheduled to take place between 2025 and 2028, will consist of major improvements to the accelerator complex, including crab cavities, and niobium and tin alloy quadrupole magnets [21]. Additionally, luminosity levelling will reduce the impact of the decrease of instantaneous luminosity during a fill. The LHC upgrades will result in an increased instantaneous luminosity of up to $7.5 \times 10^{34}\text{cm}^{-2}\text{s}^{-1}$, and an increased pile-up of 140 to 200 simultaneous interactions per bunch crossing. The ATLAS experiment is expected to collect a dataset of at least 3000 fb^{-1} (and up to 4000 fb^{-1}) during the 10 years of HL-LHC operation. The expected environment in the ATLAS detector during HL-LHC places harsh requirements on the detector. In particular, the inner detector will have to sustain high data rates and radiation [22].

For the HL-LHC, the ATLAS inner detector will be replaced with a new inner tracker (ITk), constructed entirely from silicon detectors [24]. The planned layout of the ITk detector is shown in Fig. 5.8. ITk will consist of a pixel system, closest to the beampipe, with five barrel layers and two endcap systems. The pixel detector is followed by a silicon strip detector, consisting of four barrel layers and six endcap disks. The upgraded inner detector will extend the coverage in η from $|\eta| < 2.5$ to $|\eta| < 4.0$. This thesis focuses on the upgrade of the pixel detector, which will be roughly six times larger than the current ID pixel detector. It will encompass 9400 hybrid pixel modules, with a total of 5.1 Giga-pixels and a total active area of 13 m^2. Excellent understanding of all detector components is needed before the construction of ITk, and in the following, details of the different components of ITk pixel modules are summarised.

Sensors: The pixel sensors provide the active area for detecting particles in the ITk upgrade. Sensors with a thickness of 100 or 150 μm are used, and the pixel pitch is reduced with respect to the current ID. In the innermost barrel layer, a pixel pitch of 25×100 μm^2 is used to improve the resolution in the ϕ direction, and 50×50 μm^2 pixels are used everywhere else. For the outer layers of the pixel detector *planar* sensors are used, which consist of one layer of p-type and n-type silicon, respectively. The innermost layer will use *3D sensors*, where the p- and n-type silicon is implanted as columns in the sensor, which has been shown to provide better radiation tolerance [25]. Due to the large number of sensors needed, they are produced by industry. The sensor vendors have been qualified to meet the ATLAS requirements in a dedicated Market Survey.

Readout chip: The ATLAS readout chip has been designed by the RD53 collaboration at CERN to meet the requirements on radiation tolerance and fast readout [14, 26]. The characteristics of the chip are described in detail in Sect. 6.3.

Pixel modules: The pixel sensor and front-end chips are assembled into modules through bump-bonding. The bump-bonding is also done by vendors qualified in a dedicated Market Survey. Then, the flex is glued to the assembly, and the flex and readout chip are connected using wirebonds. This step is performed by ITk institutes using varying methods. Different kinds of modules are used for the various ITk pixel layers, as summarised in Table 5.1. In the innermost layer, so-called *triplet* modules are used, which consist of three connected readout chips, each attached to one 2×2 cm^2 sensor. In the outer layer *quad* modules are used, which are four connected readout chips, attached to one 4×4 cm^2 sensor. The assembled modules will then be loaded onto local support structures to form staves for the ITk barrel or rings for the endcaps.

The ITk detector will have to withstand high radiation levels throughout its lifetime. The TID damage will be of most concern for the readout chip, affecting the logic

Table 5.1 Parameters of the ITk pixel detector layout and types of hybrid pixel modules used. Adapted from [23]

Barrel layer	Radius (mm)	Module type	Pixel pitch (μm^2)
0	34	Triplet	25×100
1	99	Quads	50×50
2	160	Quads	50×50
3	228	Quads	50×50
4	291	Quads	50×50
Endcap layer	Radius (mm)	Module type	Pixel pitch (μm^2)
0	33.20	Triplet	50×50
0.5	58.70	Triplet	50×50
1	80.00	Quads	50×50
2	154.50	Quads	50×50
3	214.55	Quads	50×50
4	274.60	Quads	50×50

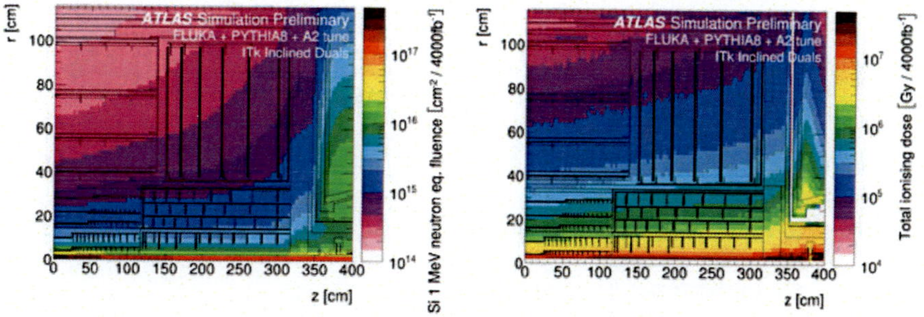

Fig. 5.9 1 MeV neutron equivalent fluence (left) and total ionising dose (right) per 4000 fb^{-1} of integrated luminosity in the ATLAS Inner Tracker. From [27]

implemented in CMOS technology, and bulk damage will be relevant for the pixel sensors, in particular leading to an increase in leakage current. The expected TID and neutron-equivalent dose for the ITk layout are shown in Fig. 5.9. The readout chip has been shown to withstand a TID dose of 1 Grad (including a safety factor), and the sensors have been characterised up to a neutron-equivalent dose of 10^{16} n_{eq}/cm^2. Therefore, the innermost pixel layers will have to be replaced after 2000 fb^{-1}, i.e. after roughly half of the lifetime of ITk.

References

1. Lutz G (2007) Semiconductor radiation detectors: device physics. https://doi.org/10.1007/978-3-540-71679-2
2. Kolanoski H, Wermes N (2016) Teilchendetektoren: Grundlagen und Anwendungen. Springer
3. Sze SM, Ng KK (2006) Physics of semiconductor devices. Wiley. ISBN: 9780470068304. https://books.google.co.uk/books?id=o4unkmHBHb8C
4. Tanabashi M et al (2018) Review of particle physics. Phys Rev D 98:030001. https://doi.org/10.1103/PhysRevD.98.030001
5. Di Saia PJ et al (2018) Clinical gynecologic oncology [electronic resource]. Eng. Ninth. Philadelphia, PA. ISBN: 9780323443166 (eBook)
6. Tsai Y (1974) Pair production and Bremsstrahlung of charged leptons. Rev Mod Phys 46:815. [Erratum: Rev Mod Phys 49:421–423 (1977)]. https://doi.org/10.1103/RevModPhys.46.815
7. **Hubbell JH, Seltzer SM (2004) Tables of X-ray mass attenuation coefficients and mass energy-absorption coefficients (version 1.4). National Institute of Standards and Technology, Gaithers-**burg, MD
8. Mayerhöfer TG, Pahlow S, Popp J (2020) The Bouguer-Beer-Lambert law: shining light on the obscure. Chem Phys Chem 21(18):2029–2046. https://doi.org/10.1002/cphc.202000464
9. p-n junction. https://en.wikipedia.org/wiki/P%E2%80%93n_junction
10. Fang J et al (2019) Understanding the average electron-hole pair-creation energy in silicon and germanium based on full-band Monte Carlo simulations. IEEE Trans Nucl Sci 66(1):444–451. https://doi.org/10.1109/TNS.2018.2879593
11. Scharf C, Klanner R (2015) Measurement of the drift velocities of electrons and holes in high-ohmic silicon. Nucl Instrum Meth A 799:81–89. ISSN: 0168-9002. https://doi.org/10.1016/j.nima.2015.07.057
12. He Z (2001) Review of the Shockley-Ramo theorem and its application in semiconductor gammaray detectors. Nucl Instrum Meth A 463(1–2):250–267. ISSN: 0168-9002. https://doi.org/10.1016/S0168-9002(01)00223-6
13. Hartmann F (2017) Evolution of silicon sensor technology in particle physics, vol 275. Springer tracts in modern physics. Springer. ISBN: 978-3-319-64434-9, 978-3-319-64436-3. https://doi.org/10.1007/978-3-319-64436-3
14. Garcia-Sciveres M, Loddo F, Christiansen J (2019) RD53B manual. Technical report CERN-RD53- PUB-19-002. Geneva: CERN. https://cds.cern.ch/record/2665301
15. Moll M (2018) Displacement damage in silicon detectors for high energy physics. IEEE Trans Nucl Sci 65(8):1561–1582. Brugger M (ed). https://doi.org/10.1109/TNS.2018.2819506
16. Faccio F, Cervelli G (2005) Radiation-induced edge effects in deep submicron CMOS transistors. IEEE Trans Nucl Sci 52(6):2413–2420. https://doi.org/10.1109/TNS.2005.860698
17. Faccio F et al (2015) Radiation-induced short channel (RISCE) and narrow channel (RINCE) effects in 65 and 130 nm MOSFETs. IEEE Trans Nucl Sci 62(6):2933–2940. https://doi.org/10.1109/TNS.2015.2492778
18. **Faccio F et al (2018) Influence of LDD spacers and H+ transport on the total-ionizing- dose** response of 65-nm MOSFETs irradiated to ultrahigh doses. IEEE Trans Nucl Sci 65(1):164–174. https://doi.org/10.1109/TNS.2017.2760629
19. Borghello G et al (2018) Dose-rate sensitivity of 65-nm MOSFETs exposed to ultrahigh doses. IEEE Trans Nucl Sci 65(8):1482–1487. https://doi.org/10.1109/TNS.2018.2828142
20. Borghello G et al (2018) Dose-rate sensitivity of 65-nm MOSFETs exposed to ultra high doses. IEEE Trans Nucl Sci 65(8):1482–1487. https://doi.org/10.1109/TNS.2018.2828142
21. Schmidt B (2016) The high-luminosity upgrade of the LHC: physics and technology challenges for the accelerator and the experiments. J Phys Conf Ser 706:022002. https://doi.org/10.1088/1742-6596/706/2/022002
22. Technical design report for the ATLAS inner tracker pixel detector. Technical report CERN-LHCC- 2017-021. ATLAS-TDR-030. Geneva: CERN (2017). https://cds.cern.ch/record/2285585

23. Expected tracking and related performance with the updated ATLAS inner tracker layout at the high-luminosity LHC. Technical report Geneva: CERN (2021). http://cds.cern.ch/record/2776651
24. ATLAS Collaboration (2017) Technical design report for the ATLAS inner tracker pixel detector. https://cds.cern.ch/record/2285585
25. Terzo S et al (2021) Novel 3D pixel sensors for the upgrade of the ATLAS inner tracker. Front Phys 9. ISSN: 2296-424X. https://doi.org/10.3389/fphy.2021.624668
26. Garcia-Sciveres M (2017) The RD53A integrated circuit. Technical report CERN-RD53-PUB-17-001. Geneva: CERN. https://cds.cern.ch/record/2287593
27. ATLAS Collaboration. Radiation simulation public results. https://twiki.cern.ch/twiki/bin/view/AtlasPublic/RadiationSimulationPublicResults

Chapter 6
Radiation Tolerance of the ITkPixV1 Pixel Readout Chip

6.1 Introduction

This chapter discusses the development and testing of the pixel readout chip for the ATLAS inner detector (ITk) upgrade. In particular, the HL-LHC will place harsh requirements on the radiation tolerance and readout speed needed for the operation of the pixel detectors. The readout chips are expected to receive total ionising doses of up to 1 Grad throughout the HL-LHC. A new pixel readout chip has been designed by the RD53 collaboration to cope with these demands [1, 2]. Here, the characterisation of the radiation tolerance of the latest readout chip prototype, ITkPixV1, is discussed. Several irradiation campaigns were performed using X-rays and protons.

First, the setup and calibration of the X-ray system in Oxford are discussed in Sect. 6.2. Then, the ITkPixV1 chip is described in Sect. 6.3, highlighting the different areas where radiation damage can be expected. The results of irradiation campaigns on bare readout chips (without sensors) are presented. Total doses of up to 2.3 Grad are reached, at a dose rate of 4 Mrad/h. Different aspects of the chip are studied: the digital logic (Sect. 6.4), the analogue front-end (Sect. 6.5), and the dose rate dependence of the damage for both parts of the chip, as well as the powering circuit (Sect. 6.6). Finally, the results of an irradiation programme using protons are presented in Sect. 6.7. Here, the focus is the performance of an ITkPixV1 module, especially the impact of the radiation damage in both the readout chip and the sensor.

6.2 Calibration of the OPMD X-ray System

Several X-ray irradiation campaigns of ITkPixV1 readout chips were performed using an X-ray system at the Oxford Physics Microstructure Detector (OPMD) facility. The following section describes the general concepts of X-ray production and the calibration of the X-ray system in OPMD.

M. Mironova, *Search for Higgs Boson Decays to Charm Quarks with the ATLAS Experiment and Development of Novel Silicon Pixel Detectors*, Springer Theses,
https://doi.org/10.1007/978-3-031-36220-0_6

6.2.1 X-ray Production and Fluorescence

X-rays are typically produced by accelerating electrons (released from a hot cathode) and colliding them with a metal target (anode). The electrons can undergo two different processes within the anode, which contribute to the X-ray spectrum.

Bremsstrahlung Bremsstrahlung describes the process of scattering electrons in the strong electric field near a nucleus with a high atomic number Z. The deceleration of the electrons causes an X-ray photon to be emitted, with its energy equal to the lost kinetic energy of the electron. In this way, photons with a continuous spectrum of energies are produced, only limited by the energy of the incoming electrons.

Characteristic radiation If an electron has enough energy, it can eject an orbital electron from the inner shell of an atom. Then, an electron from an outer shell can fill the vacancy while emitting an X-ray photon. Photons are produced at discrete energies, defined by the initial and final shell of the electron. Each atomic shell is defined by its principal quantum number n (n = 1,2,3...) and labelled with a corresponding letter (K, L, M). The X-ray lines are named based on which shell the electron moves to and where it comes from. For example, a K_α photon is emitted when the electron moves from the L to the K shell, and a K_β photon is emitted when it moves from the M to the K shell. The K_α line usually has the highest intensity.

X-rays are usually produced in X-ray tubes, with a schematic of such a tube shown in Fig. 6.1. The cathode produces the electrons, which are then accelerated towards the anode by applying a high voltage, called tube voltage. An electric current is created, referred to as the tube current. Electrons hit the target, usually made out of copper, tungsten or molybdenum, and produce X-rays that exit the tube through a small exit window, which commonly has a filter to remove low-energy components. The X-ray spectrum depends on the acceleration voltage, i.e. the electron energy,

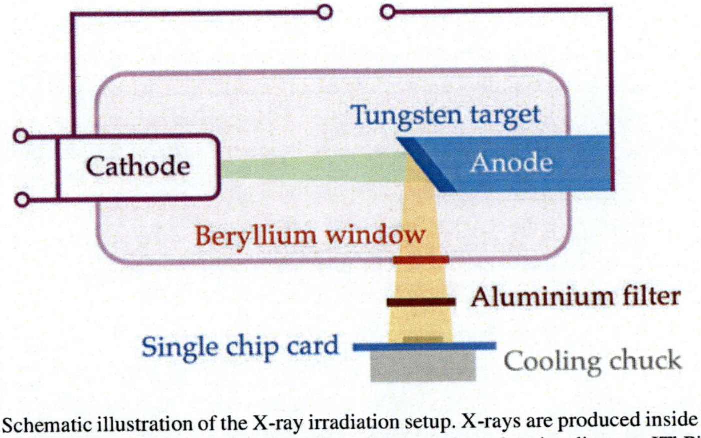

Fig. 6.1 Schematic illustration of the X-ray irradiation setup. X-rays are produced inside the X-ray tube, passed through Aluminium and Beryllium filters, and used to irradiate an ITkPixV1 chip mounted on a cold chuck

and the anode material. X-ray production generates a significant amount of heat, so the target must be cooled, e.g. with a water-cooled chuck placed behind the anode.

6.2.2 *Characterisation of the Oxford X-ray Irradiation Setup*

This section describes the setup and calibration of the new X-ray system in the Oxford Physics Microstructure Detector facility (OPMD) [3]. A picture of the X-ray system is shown in Fig. 6.2 (left). The X-rays are produced using a tungsten target and filtered by a 0.8 mm Beryllium window at the tube exit. The tube can be operated with an anode voltage of 5 kV to 60 kV and a current between 1 and 60 mA, with the maximum achievable power being 3 kW.

As a first step, the spectrum of the X-ray tube was measured using a silicon drift detector. Silicon drift detectors measure the energy of a photon depending on how much ionisation it produces and are known for having excellent energy resolution. The detector was calibrated using an ^{55}Fe source, which has well-defined $K_{\alpha 1/2}$ lines around 5.9 keV. The silicon drift detector was then placed in the X-ray system. To not saturate the detector, it was then positioned as far away from the aperture of the X-ray tube as possible. The energy spectrum was measured as a function of anode voltage while keeping the X-ray tube current low. An example of the energy spectrum at an anode voltage of 40 kV is shown in Fig. 6.2 (right). The continuous Bremsstrahlung spectrum is visible and peaked at approximately 9 keV. There are also various characteristic lines visible, with their origins summarised in Table 6.1, based on [4]. As expected, many lines can be identified to originate from the tungsten target. However, surprisingly, the most prominent characteristic lines stem from

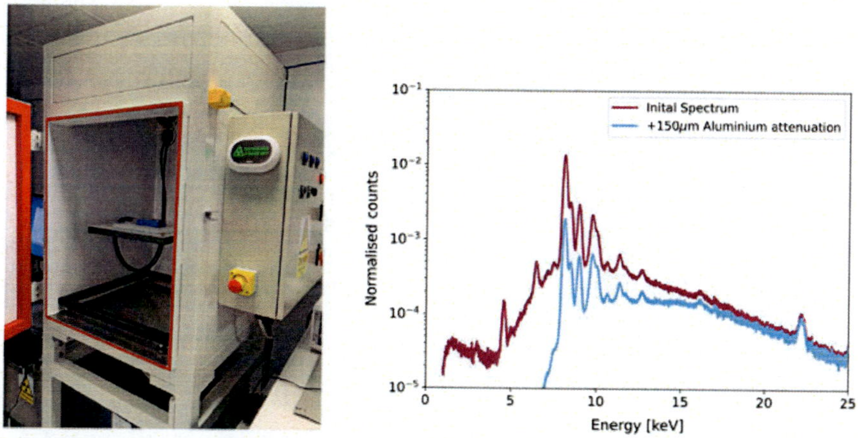

Fig. 6.2 Image of the Oxford X-ray setup (left) and the energy spectrum of the X-ray tube, measured using a silicon drift detector (right)

Table 6.1 The characteristic X-ray lines identified in the measured X-ray spectrum in Fig. 6.2 (right), based on data from [4]

Energy	X-ray line	Energy	X-ray line
4.5 keV	Ti $K_{\alpha 1/2}$	9.0 keV	Cu $K_{\beta 1}$
6.4 keV	Fe $K_{\alpha 1/2}$	9.7 keV	W $L_{\beta 1}$
8.1 keV	Cu $K_{\alpha 1/2}$	10.0 keV	W $L_{\beta 2}$
8.4 keV	W L_{α}	11.3 keV	W $L_{\gamma 2}$

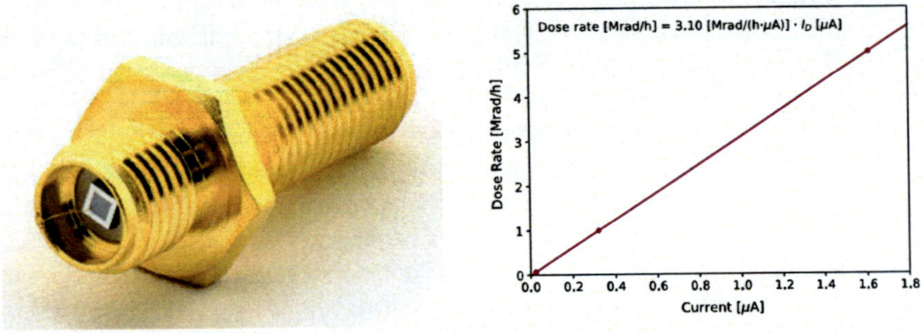

Fig. 6.3 Picture of the AXUVHS5 diode (left), which was used for the calibration of the X-ray setup, from [5], and dose rate as a function of measured diode current (right). Several points indicate examples of the dose rates used in irradiation campaigns

copper, at 8.1 and 9.0 keV. Behind the tungsten target, there is a copper chuck, which provides the cooling of the target, so these lines are likely caused by electrons producing characteristic X-ray radiation there. For the irradiation of the ITkPixV1 readout chips, the X-rays are additionally filtered using a 150 μm Aluminium filter to remove low energy components, with the resulting spectrum also shown in Fig. 6.2 (right). In the subsequent dose characterisation measurements, the Aluminium filter is always included.

The delivered dose of the X-ray set was measured as a function of various parameters of the setup using a calibrated diode. The diode, shown in Fig. 6.3 (left), has a sensitive depth of 50 μm and a sensitive area of 1×1 mm^2. Several diodes were calibrated at CERN [5] to ensure that the setups used in the ITkPixV1 irradiations share uniform calibrations. These diodes were calibrated using an X-ray tube with a tungsten target, a 150 μm Beryllium window, and a 150 μm Aluminium filter. The X-ray tube was operated at 40 kV. Consequently, the conditions in the Oxford setup were matched to this as closely as possible.

The dose rate was determined by measuring the diode current when biasing the diode to –50 V, and subtracting the dark current without irradiation. This current can then be converted to SiO_2 equivalent dose rate using the calibration curve which is available for each diode, with an example shown in Fig. 6.3 (right). The typical dose rates used for ITkPixV1 irradiations are indicated on the plot.

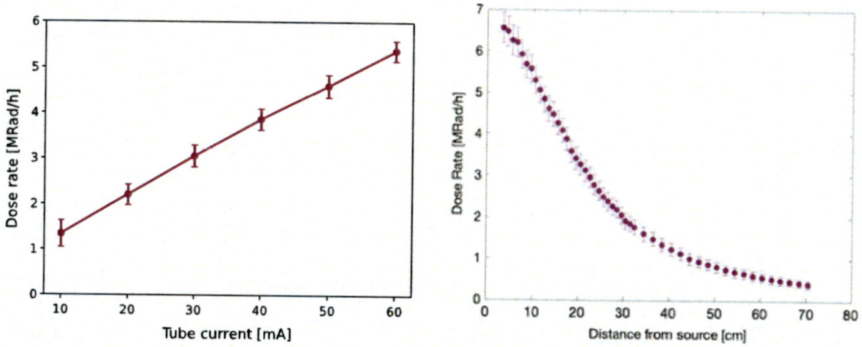

Fig. 6.4 Dose rate as a function of X-ray tube current (left) and as a function of distance from the source (right)

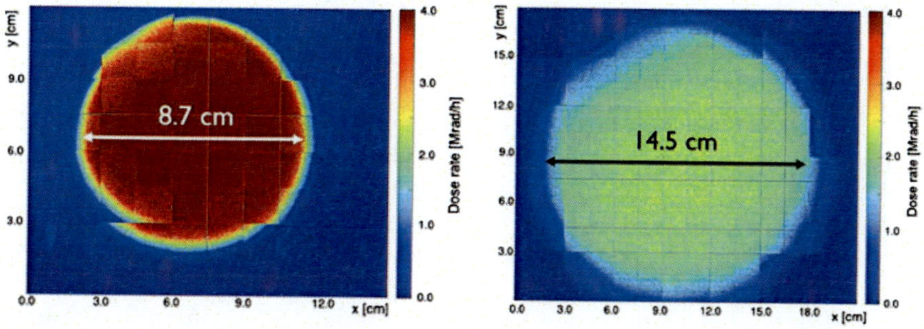

Fig. 6.5 xy-profiles of the beam obtained in the calibration of the X-ray system at a distance of 10 cm (left) and 20 cm (right) from the source

As the first calibration measurement, the dose rate was measured as a function of X-ray tube current. This is shown in Fig. 6.4 for an operational voltage of 40 kV, with the diode placed at roughly 10 cm from the exit of the X-ray tube. A linear behaviour of the intensity, and thus dose rate, with tube current is expected and observed. Next, the dose rate as a function of distance from the source was measured. A roughly $1/r^2$ dependence can be observed, as shown in Fig. 6.4.

Finally, the beam spot of the X-ray tube was measured at two different distances from the aperture of the tube. This was done with a Medipix Merlin detector, a commercially available pixel detector with excellent time resolution, with frame rates over 1 kHz, which can be used to count photon hits per period of time [6]. In this case, the detector was used to count the number of photons per 10 μs window, at a tube voltage of 40 kV. The detector has a size of approximately 4×4 cm^2 and is made of four single chips. It was moved by hand in a grid to cover the area. The stitched images of the beam spot at a distance of 10–20 cm from the beam spot are shown in Fig. 6.5, with the approximate diameters of the beam spot noted on the plots. Overall, the beam spot is very circular and uniform, with some loss in intensity at the edges.

6.3 The ITkPixV1 Chip

The ITkPixV1 chip is the prototype readout chip for the ATLAS ITk upgrade, using a design framework called RD53B with 65 nm feature size CMOS technology and developed in collaboration with CMS [2]. Its predecessor, the RD53A chip [1], has been tested and characterised extensively. RD53B is a generic pixel readout chip framework that can be used for chips of different sizes. It consists of a *pixel matrix* and a *chip bottom*. The pixel matrix is built up of identical 8 × 8 pixel cores stepped and repeated in columns and rows. A core is physically 400 μm × 400 μm. The chip bottom contains all of the system functionality. The ATLAS version of the RD53B chip, ITkPixV1, has 48 core rows and 50 core columns, 50 μm × 50 μm pixel pitch, and a total of 400 × 384 pixels. The total size of the chip is 20 × 21 mm^2.

6.3.1 *Powering of the ITkPixV1 Chip*

The ITk Pixel modules will be powered in serial power mode, to reduce the required number of cables and the detector material significantly. The ITkPixV1 chip powering is implemented using Shunt LDO regulators (SLDO) [7]. The SLDO combines a linear Low Drop-Out voltage regulator (LDO) and a shunt element. Such a setup can provide constant current operation with multiple devices connected in parallel, which is not possible using a simple LDO. The SLDO circuit was first used in the FE-I4 readout chips used in IBL, and the design has significantly evolved up to the final circuit used in ITkPixV1.

The basic principle of the SLDO is illustrated in Fig. 6.6. A standard LDO is used to power the main load, L, and an internal load L_s in parallel. The latter is the shunt element and it is controlled to achieve the desired current at the input, regardless of the main load L. The system is designed to operate chips at a constant current, i.e. $I_{in} = I_L + I_s =$ constant. This current must exceed the maximum current required by any chip at any moment, and account for chip-to-chip variations. In practice, the several chips in a module are powered in parallel and the current sharing is set by the resistive behaviour of the regulators, which can be described by:

$$I_{in} = I_L + I_s = \frac{V_{in} - V_0}{R_{eff}} \qquad [V_{in} > V_0] \tag{6.1}$$

where V_0 is a constant offset and R_{eff} provides the ohmic behaviour needed to share current between multiple devices. A diagram of this behaviour is shown in Fig. 6.6 (right), with the voltage shown as a function of input current. Such V-I curves can also be measured experimentally to characterise the SLDO behaviour in the chip.

A schematic of the actual circuit used in ITkPixV1 is shown in Fig. 6.7. The LDO is highlighted in red, and the blue part provides the offset V_0, which is controlled by a reference voltage labelled V_{offset}. The main load L is the chip itself and is not shown

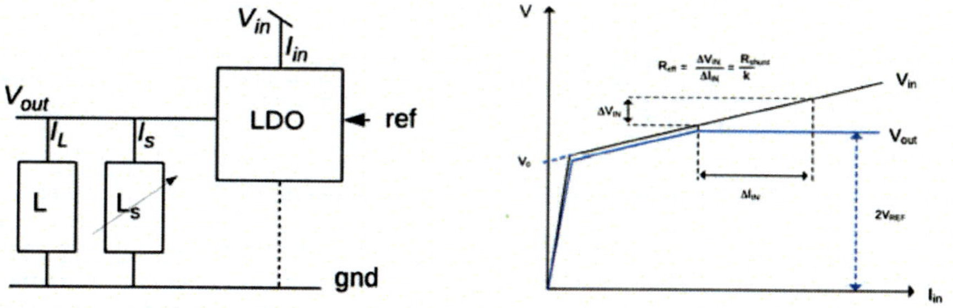

Fig. 6.6 Illustration of SLDO circuit in RD53B. Concept of SLDO operation as a linear regulator (LDO) powering an main load L and a variable internal shunt load L_s (left). Desired current versus voltage characteristics for SLDO (right). The unregulated input voltage and regulated output voltage are shown. From [2]

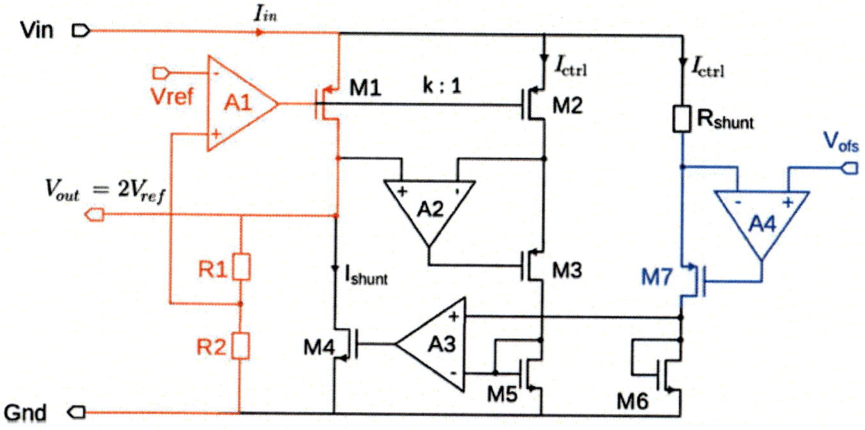

Fig. 6.7 Simplified schematic of the RD53B SLDO regulator. The colors differentiate the LDO (red), shunt (black), and offset (blue) functionality. From [2]

in the diagram, and the device M4 is the internal load, with the rest of the circuitry (indicated in black) being the active control. The SLDO circuit is designed to be compatible with a 2 V input voltage and includes an over-voltage and under-shunt transient protection.

To operate the SLDO in a serial power chain, the SLDO has to become active immediately upon current flow, work with high efficiency and uniformity and run for a wide range of temperatures. In order to meet these conditions, the power-up scheme was significantly improved compared to what was previously implemented in RD53A. ITkPixV1 does not use the SLDO output to power any reference circuit, e.g. the one necessary for providing the reference current in the chip (I_{ref}). Instead, an additional low current linear regulator (the *preregulator*) is used to power the main reference current generator. Then, all other reference currents are derived from this main reference current. The I_{ref} is generated by the Core Bandgap and can be adjusted by wire bond settings to compensate for process variations [2]. The nominal value of the reference current is 20 μA. From this reference current, the two LDO reference voltages are generated for the analogue and digital domains: `VrefA` and `VrefD`, which then influence the voltages powering the analogue and digital domain: `VDDD` and `VDDA`. In summary, the following voltages and currents are relevant for the ITkPixV1 chip operation:

- `VinA, VinD`: Input voltages to the LDO regulators for the analogue/digital domains
- `VrefA, VrefD`: Reference voltages for the analogue/digital voltage, generated from the main reference current
- `VDDA, VDDD`: Voltage supplied by the LDO to the analogue/digital part of chip
- `V_offset`: Offset voltage of the SLDO circuit
- `VDD_PRE`: Reference voltage for the preregulator circuit
- `Vref_ADC`: Reference voltage for the ADC
- `Iref`: Main reference current
- `I_in analogue/digital`: Input current $I_{in} = I_L + I_s$ for the analogue and digital domain
- `I_shunt analogue/digital`: Shunt current for the analogue and digital domain.

6.3.2 Analogue Front-End

The ITkPixV1 uses a differential analogue front-end, shown in Fig. 6.8. In brief, the analogue front-end functionality is as follows. The circuit receives a current pulse, either from the sensor or the injection circuit. This pulse is then fed into a charge sensitive amplifier, which produces a triangular pulse as a function of time. A comparator compares this pulse to a threshold voltage value. The time the signal is above the threshold, Time-over-Threshold (*ToT*), is measured and transferred into the digital power domain, where it is converted into a digital signal and further processed.

The *pre-amplifier (preamp)* is the first stage of this circuit. It has a straight regulated cascode architecture with NMOS input transistor in weak inversion. A high-VT (high threshold voltage) type input device is used. It has a continuous reset, and its gain can be adjusted by adding a feedback capacitor `Cf0`. The preamp has two main biases: the main bias (input transistor current), set by the `DiffPreamp`, and the continuous reset feedback current, set by the `DiffVff`. For the operation on the chip in ITk, different settings of these parameters are considered, depending on the layer in the detector and the expected data rates. The settings vary from the inner layers (`DiffPreamp` = 900 and `DiffVff` = 150) to the outer layers (`DiffPreamp` = 400 and `DiffVff` = 60). A leakage current compensation circuit provides (LCC) an option for additional feedback in the preamp. It is an active low-pass filter, and the bias of the LCC can be globally adjusted using the `DiffLcc` setting. The LCC should only be used in the case of a large sensor leakage current (i.e. 2nA or higher),

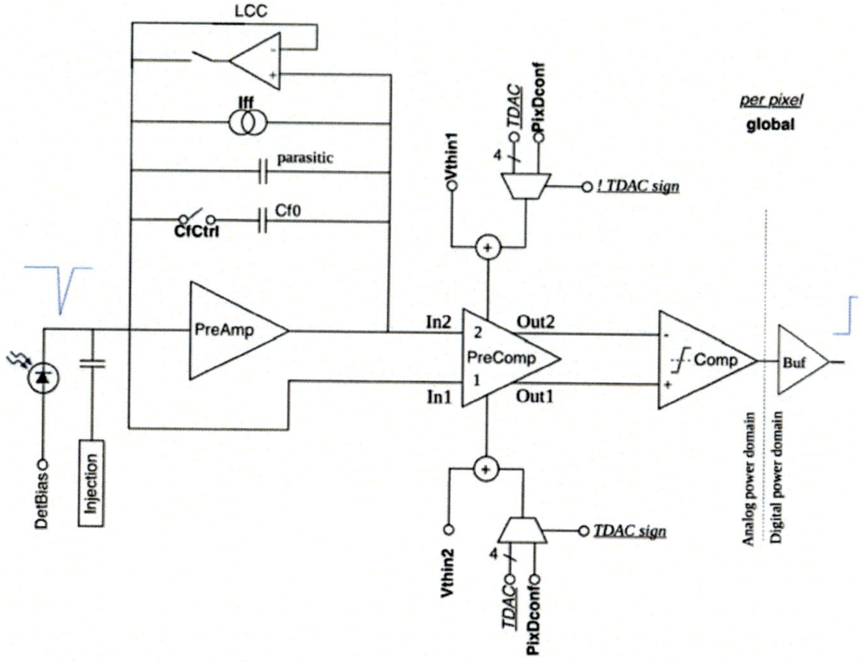

Fig. 6.8 Schematic of ATLAS chip analogue front end with differential second stage and comparator. The underlined italic signals are controlled per pixel, while the boldface signals are global. A digital buffer is manually placed at the comparator output. From [2]

for example, after irradiation. With increasing leakage current, an increase in LCC bias is recommended. Input and output of the preamp are taken as a differential input to the next stage.

The second stage of the analogue front-end is the *pre-comparator* and its bias current can be globally adjusted using the `DiffPreComp` setting. The primary purpose of the pre-comparator is to introduce the threshold. The global threshold of all pixels can be adjusted through two distributed voltages (`Vth1` and `Vth2`) which introduce an offset between the two branches of the pre-comparator. In each pixel, the threshold can be trimmed using one 4-bit resistor ladder in each pre-comparator branch, which can be modified using the `TDAC` setting.

Finally, the pre-comparator is followed by a classic continuous time comparator stage (referred to as *comparator*). The bias current of the comparator is adjustable through the `DiffComp` register. The output of the comparator is connected to the digital pixel region through a buffer placed near the comparator output.

The settings relevant to ITkPixV1 operation and their typical values are summarised in Table 6.2.

Table 6.2 Parameters and typical settings available in the analogue front-end of ITkPixV1

Setting	Description	Usual setting [DAC units]
`DiffPreAmp`	Main preamp bias current	900 or 400
`DiffVff`	Continuous preamp reset feedback current	150 or 60
`DiffLccEn`	Leakage current compensation enable bit	0 or 1
`DiffLccEn`	Leakage current compensation bias	0–800
`DiffPreComp`	Main pre-comparator bias current	300
`Vth1`	Main threshold voltage	Tuned
`Vth2`	Secondary threshold voltage	50
`TDAC`	Pixel threshold trim-bit	Tuned
`DiffComp`	Main comparator bias current	500

6.3.3 Chip Readout

The ITkPixV1 chips can be tested independently or assembled into modules to test together with a sensor, either as single or quad modules. For bench testing, and in the studies shown in the following, single chips or modules are used, which are mounted onto single chip cards (SCCs). These are PCBs made for testing, with dedicated pins for monitoring various chip properties, such as chip voltages. A picture of an SCC is shown in Fig. 6.9 (left).

The YARR readout system is used for the chip readout [8]. YARR is a PCI-based readout system developed for ATLAS readout chips, used for the FE-I4, RD53A and now ITkPixV1 chips. The chip is connected to the readout system using a DisplayPort cable, and an example of a setup is shown in Fig. 6.9 (right). Using YARR, it is

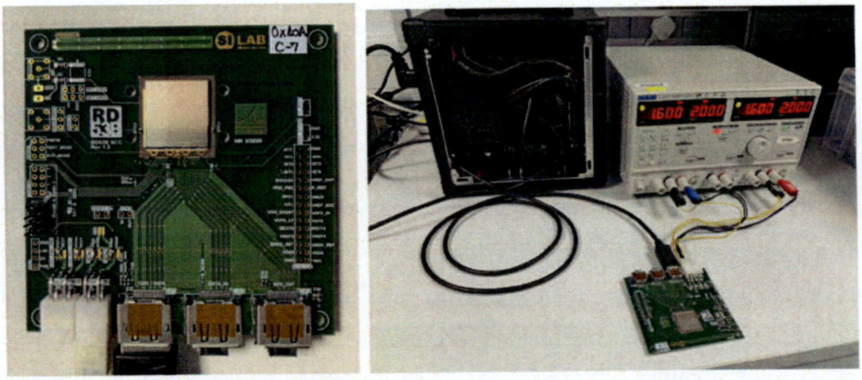

Fig. 6.9 Pictures of the ITkPixV1 single chip card (left) and testing setup using the YARR readout system (right)

possible to configure the various chip settings to tune and operate the chip. In particular, the readout system is used to characterise the chips with various measurements (*scans*).

Digital Scan

In a digital scan, a configurable amount of charge is injected into the digital domain of the pixel several times. The chip then counts the number of hits it sees and reports the occupancy. Ideally, each pixel should see the exact number of pulses injected. This scan is usually run first to check chip communication and ensure good operation of the digital circuitry.

Analogue Scan

Like a digital scan, an analogue scan injects a certain number of pulses into the analogue front-end and counts the number of hits received. Ideally, the number of injected and measured pulses is the same, but the values are more likely to differ in an analogue scan due to noise in the analogue front-end. Additionally to the occupancy of the device, the scan also provides information about the mean time-over-threshold (*ToT*) and time-of-arrival (*ToA*) for each pixel. The analogue scan is usually used to validate the performance of the analogue front-end.

Threshold Tuning

Different scans are available for adjusting the global and pixel threshold values, but the general principle is the same. A charge with the value of the desired threshold is injected in the analogue front-end, and a global or pixel chip threshold parameter is varied (i.e. scanning `Vth1` from high to low values) until 50% of the pixels register a hit. This optimisation can be performed for the global parameter or each pixel individually. Typical threshold values are of the order of 1000 to 2000 electrons.

Threshold Scan

Finally, it is possible to measure the resulting distribution of pixel thresholds using a threshold scan. Here, a charge is injected in each pixel several times, and the response of each pixel is measured. This is repeated for a range of increasing charges. The occupancy will turn on as a function of injected charge for each pixel, resulting in a so-called S-curve. The 50% occupancy point for each pixel is defined as the pixel threshold, and the noise is derived from the width of the S-curve. From this, the overall distribution of the threshold and noise in the chip can be derived.

6.4 Damage to the Digital Logic

The first tests of the radiation tolerance of the ITkPixV1 were performed to characterise the damage to the digital logic. Several dedicated X-ray irradiation campaigns were conducted.

6.4.1 *Ring Oscillators*

The ITkPixV1 chip includes several test circuits, called ring oscillators, which can be used to characterise the radiation damage to different gates used in the digital chip logic. These ring oscillators are made with varying types of logic cells and various sizes of gates. Each ring oscillator drives a 12-bit counter, which is enabled for a given time. During this time, the ring oscillator output oscillates between 0 and 1 with a particular frequency, given by the delay of the gate used. This is illustrated in Fig. 6.10.

In the ITkPixV1 chip, there are a total of 42 ring oscillators located in different groups at the bottom of the chip. Table 6.3 summarises the different kinds of ring oscillators in ITkPixV1, their transistor drive strength, the length of the oscillators (i.e. the number of gates in each circuit) and the delay measured before irradiation. The transistor drive strength is related to the size of the transistor and proportional to W/L, where W and L are the width and length of the CMOS transistor, respectively. Therefore, the strength 0 gates are the narrowest transistors, and are expected to suffer the most from radiation damage [9]. The strength 1 and strength 4 gates represent most closely what is used to implement the digital logic of the chip. The ring oscillators are grouped in different *banks*, and the Bank A ring oscillators correspond exactly the ring oscillators used in the previous RD53A chip [1], to allow for a direct comparison.

For each ring oscillator, it is possible to calculate the oscillation frequency f, or the individual gate delay $T = 1/(N \times f)$, where N is the length of the ring oscillator. The gate delay will increase with increasing radiation damage, and the ring oscillator frequency will decrease, which is the main measure of radiation damage to the digital logic. From simulations of the ITkPixV1 chip, it is expected that the digital logic

Fig. 6.10 Schematic of a ring oscillator block (left), from [2], and illustration of the ring oscillator output (right)

Table 6.3 Ring oscillators available in the ITkPixV1 chip, using different gates, drive strength, length (in number of gates) and examples of measured gate delay on an ITkPixV1 chip

Ring oscillator number	Type	Strength	Length	Approx. Delay [ns]
Bank A				
0	Inv. CLK driver	0	55	0.042
1	Inv. CLK driver	4	51	0.029
2	Inverter	0	55	0.044
3	Inverter	4	55	0.029
4	4-input NAND	0	19	0.129
5	4-input NAND	4	19	0.076
6	4-input NOR	0	19	0.113
7	4-input NOR	4	19	0.076
Bank B left/right				
0 & 1	Inv. CLK driver	0	38.2	0.040
2 & 3	Inv. CLK driver	4	44.5	0.029
4 & 5	Inverter	0	38.1	0.041
6 & 7	Inverter	4	44.3	0.030
8 & 9	4-input NAND	0	12.6	0.120
10 & 11	4-input NAND	4	16	0.079
12 & 13	4-input NOR	0	14.5	0.110
14 & 15	4-input NOR	4	14.5	0.086
Bank B FF				
16 & 17	Scan D-flip-flop	0	6.1	0.372
18 & 19	D-flip-flop	1	6.2	0.308
20 & 21	Neg. edge D-flip-flop	1	5	0.375
Bank B LVT				
22	LVT inverter	0	40.6	0.031
23	LVT inverter	4	56	0.021
24	LVT 4-input NAND	0	16.5	0.087
25	LVT 4-input NAND	4	22.8	0.059
Bank B CAPA				
26–33	Inj-cap loaded 4-input NAND	4	16.8	0.078

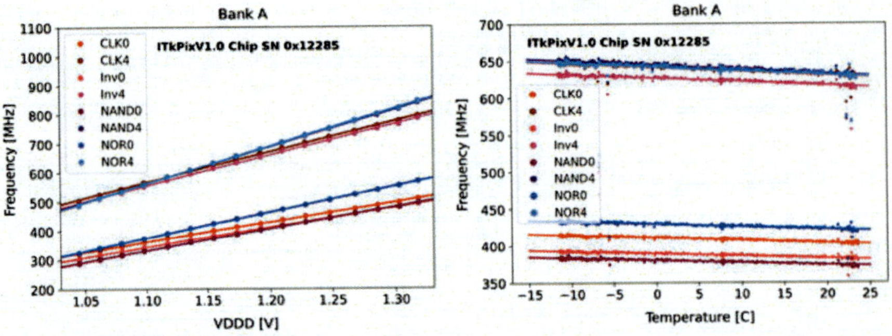

Fig. 6.11 Examples of ring oscillator calibrations performed on ITkPixV1 chip before irradiation. Ring oscillator frequencies are shown as a function of VDDD voltage (left) and temperature (right)

will start to fail at a delay increase of more than 200% from its original value. The gate delays for the various gates before irradiation vary between 0.03 and 0.4 ns.

Before performing irradiations, some preliminary measurements have to be taken. There is a strong dependence of ring oscillator frequency on the supplied voltage and a small dependence on temperature. These dependencies were characterised for each chip before irradiation.

First, the frequency of the ring oscillators was measured while varying the VDDD voltage. The result is shown in Fig. 6.11 for a subset of the ring oscillators. A strong linear behaviour is visible, and the ring oscillator frequency increases by up to 300 MHz when increasing VDDD from 1.0 to 1.3 V. The slopes and offsets are determined for each ring oscillator. These parameters are then used to account for any changes in VDDD during the irradiation.

Next, the dependence of the ring oscillators on temperature was measured. This was done by slowly cooling down the setup from room temperature to −12 °C and monitoring the ring oscillator frequency while accounting for any changes in VDDD using the parameterisations derived above. The results are also shown in Fig. 6.11. A clear but small linear frequency dependence on the temperature can be seen. The frequency changes by up to 20 MHz when changing the temperature from −12 °C to room temperature.

6.4.2 *Irradiation Setup*

All irradiations focused on the ring oscillators were performed using a similar methodology. The ITkPixV1 chips were irradiated in the X-ray setup described in Sect. 6.2. The X-ray tube was operated at a voltage of 40 kV, with a 150 μm Aluminium filter placed at the tube exit. The desired dose rate was achieved by adjusting the tube current and placing the chip at a certain distance from the tube exit.

The irradiations were performed while the chip was cooled. This was achieved by placing the chip on a chuck connected to a Huber CC-508 chiller [10], which was able to keep the chip at a constant temperature of roughly $-10\,°\mathrm{C}$ ($\pm$ 1 °C). This corresponds to a chip operating temperature of approximately $-7.5\,°\mathrm{C}$, which is the expected maximum temperature during ITk operation, and more radiation damage is expected at higher temperatures. The temperature was monitored by reading the resistance of an NTC (negative temperature coefficient) thermistor, located on the single-chip card [11]. The chip was powered by supplying 1.6 V to the chip. This voltage is then distributed to the analogue and digital domains of the chip via the LDOs, leading to a nominal 1.2 V supplied on each. The voltage provided to the digital domain (VDDD) was monitored. The primary measurement during irradiations was the frequency of the ring oscillators. The ring oscillators were read out at least after every 0.1 Mrad of delivered dose. While the ring oscillators are not measured, the chip was kept busy by running basic digital scans to avoid the chip sitting in an idle state and simulate more precisely what would happen to a chip operating in the HL-LHC environment.

Irradiations were performed at different dose rates. First, irradiations at high dose rates (around 1–4 Mrad/h) were performed, and the results are described in Sect. 6.4.3. Then, the dependence of the radiation damage on the dose rate was studied, with the results detailed in Sect. 6.4.4. The quoted dose rate is reduced by 20%, compared to the dose rate measured by the calibrated diode due to attenuation from a thin copper layer covering the ring oscillators in the chip.

6.4.3 High Dose Rate Irradiation up to 1 Grad

The expected delivered dose to the innermost layers of ITk after $2000\,\mathrm{fb}^{-1}$ of luminosity is 1 Grad, after which these layers will be replaced. Consequently, several irradiation campaigns up to a TID of 1 Grad were performed, and the first of these irradiation campaigns is described here in more detail.

The ring oscillator frequencies and the monitored environmental parameters are shown in Fig. 6.12 (left). The temperature stayed within $-10\,°\mathrm{C}\ +/-\ 1\,°\mathrm{C}$. The VDDD voltage increased continuously during irradiation and was reset to 1.2 V multiple times. The ring oscillator frequencies have been adjusted for the change in VDDD, according to the calibration measurements. The evolution in ring oscillator frequencies shows a smooth decrease. After 400 Mrad, there is an increase in noise on the ring oscillator frequencies, which is caused by additional measurements taken during this period, which increase the variations in VDDD voltage.

It is also possible to use the ring oscillator frequencies to calculate the relative increase in gate delay, as $\Delta T = 1/(N \times f) - 1/(N \times f_{\mathrm{start}})$, where f_{start} is the ring oscillator frequency before irradiation. The relative gate delay increases for the Bank A ring oscillators in ITkPixV1 are shown in Fig. 6.12 (right). Similar irradiation campaigns have been performed for three additional ITkPixV1 chips, and the measured increase in gate delay at 1 Grad is summarised in Fig. 6.13 for all gates and irradi-

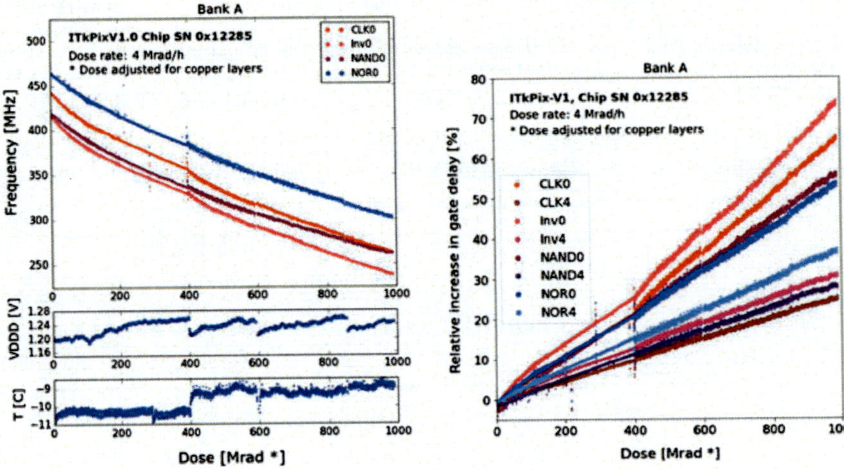

Fig. 6.12 Ring oscillator frequencies for the strength 0 gates for an irradiation up to 1 Grad, at a dose rate of 4 Mrad/h (left). The bottom panels show the evolution of the VDDD voltage and temperature during irradiation. Relative increase in gate delay with dose for the Bank A ring oscillators in ITkPixV1 (right)

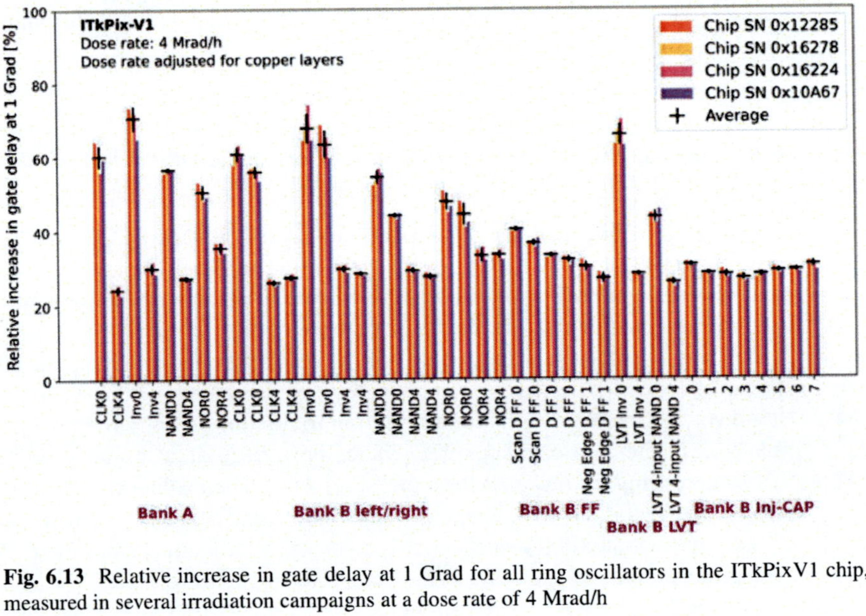

Fig. 6.13 Relative increase in gate delay at 1 Grad for all ring oscillators in the ITkPixV1 chip, measured in several irradiation campaigns at a dose rate of 4 Mrad/h

ation campaigns. The results of the individual campaigns are shown, as well as the average and standard deviation of the measured values. For the Bank A and Bank B ring oscillators of similar types, the results are very consistent, and the gate delay increase at 1 Grad is of the order of 70% for strength 0 gates and of the order of 30% for strength 4 gates. As expected from single transistor measurements [12], the delays of the narrower strength 0 gates increase more rapidly than those of the larger strength 4 gates. For the flip-flop ring oscillators, both the strength 0 and strength 1 gates show delay increases of the order of 40%. The behaviour of the injection capacitor loaded NAND gates all show good agreement, with an increase of gate delay between 25 and 30% at 1 Grad. These gates allow understanding the spread of the irradiation results and the variability of the injection capacitor values. Overall, for these irradiation campaigns at high dose rates, the resulting relative increase in gate delay is well within the expected margin of <200% for all the ring oscillators in ITkPixV1.

6.4.4 Measurement of Dose Rate Dependence

The primary quantity of interest when considering the ring oscillators and the digital logic is the increase in relative gate delay at the end of life, which should be less than 200% after 1 Grad to ensure full functionality. While the high dose rate (HDR) irradiations show promising results, there is a further complication to consider in understanding the expected radiation damage to the chip at the end of life in the HL-LHC. In ITk, the total dose of 1 Grad will be delivered over many years rather than over a few weeks. In small transistors, like the ones used to implement the digital logic in ITkPixV1, radiation damage has been observed to strongly depend on the dose rate, with a low dose rate (LDR) causing more damage [12]. Thus, it is essential to understand the behaviour of the chip in low dose rate irradiations and the dose rate dependence of the radiation damage for the ITkPixV1 chip. Several irradiation campaigns were performed at lower dose rates: 20, 40, 60 and 80 krad/h. These irradiation campaigns were conducted over several weeks up to a TID of 10–20 Mrad. These doses are significantly lower than those reached in high dose rate irradiations. An additional irradiation campaign at a very high dose rate of 27 Mrad/h was also performed by placing the chip closer to the exit of the X-ray tube and centring the ring oscillators in the beam spot, such that they receive the highest possible dose.

Examples of the relative gate delay increase at different dose rates are shown in Fig. 6.14. There is a clear difference between the damage at different dose rates, and the gate delay increases significantly quicker for low dose rates. The effect seems to be more dramatic for the strength 0 gates. For each ring oscillator, the slope of gate delay increase is determined by fitting the linear part of the delay increase, starting at approximately 5 Mrad. The results of the linear fits are also shown in Fig. 6.14. Figure 6.15 shows the fitted slopes as a function of dose rate for each type of ring oscillator gate. The top panel in each plot shows the fitted values, and the bottom

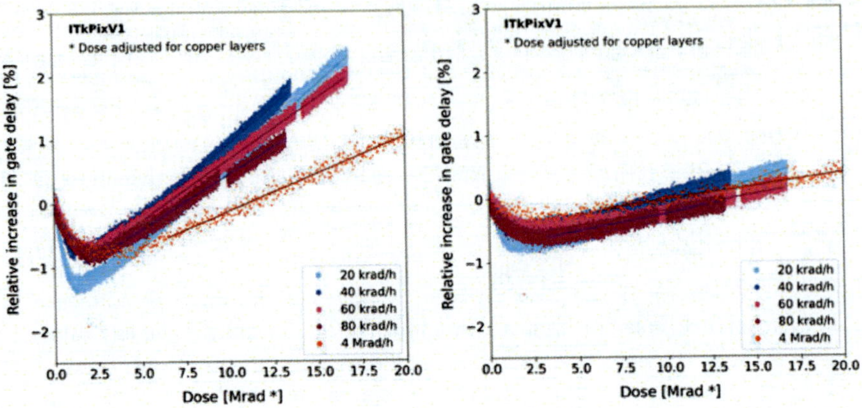

Fig. 6.14 Gate delay increases at different dose rates for the ring oscillator made of inverter gates, for the strength 0 (left) and strength 4 (right) gates

panel shows the ratio of the fitted slopes compared to the result obtained in the high dose rate irradiation at the most frequently used dose rate of 4 Mrad/h. The results are averaged between the Bank A and Bank B ring oscillators of the same type. A systematic uncertainty on the dose rate is considered, depending on the precision with which it could be measured using the calibrated diode and the accuracy of the chip position in the beam spot. This uncertainty is usually of the order of 5–10%. A systematic uncertainty on the fitted ring oscillator delays was also considered by varying the range fitted and the standard deviation of the results in the case of averaged values.

There is a clear trend of increased radiation damage at low dose rates. The increase in damage at low dose rates depends on the transistor size, with the strength 0 gates receiving roughly three times more damage at low dose rate and the strength 4 gates receiving approximately 2.5 times more damage. However, the conversion factor also depends on the exact type of gate used.

Considering these conversion factors, the expected damage after a lifetime in HL-LHC at low dose rate can be estimated. For the strength 4 gates, which are representative of the chip digital logic, the expected delay increases up to 90% at low dose rates, which is well within the requirements. It is also worth noting that the behaviour of growing damage seems to flatten out at low dose rates, which is consistent with what was observed in single transistor measurements [12]. This suggests that there will not be significantly more damage to the digital logic gates at even lower dose rates, e.g. in the outer layers of ITk.

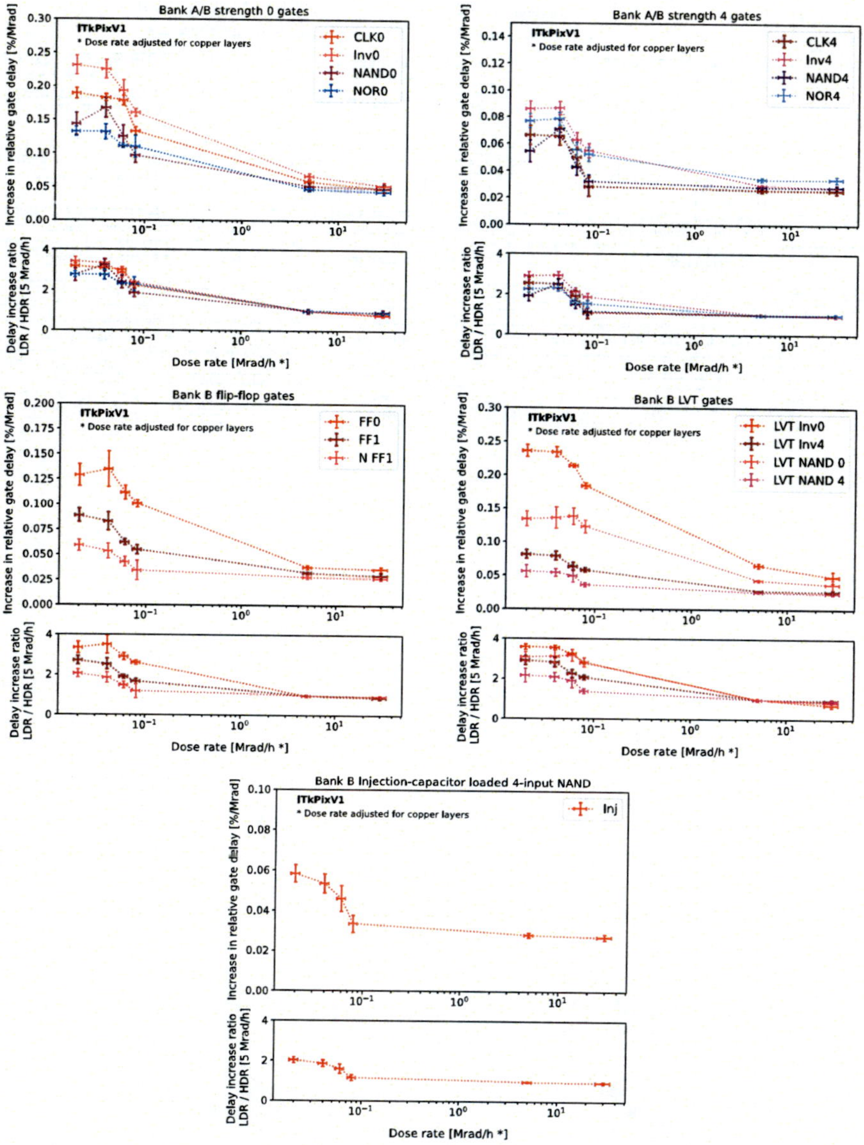

Fig. 6.15 Fitted increase in relative gate delay as a function of dose rate for all types of gates used in the ITkPixV1 ring oscillators. The points correspond to the slopes derived in the linear parts of the curves shown in Fig. 6.14

6.4.5 *Comparison of X-ray and Kr-85 Irradiations*

The low dose rate irradiation studies described above have only been performed at low TID. To fully understand the behaviour of the ITkPixV1 chip in the HL-LHC environment, studies at high TID are also necessary at low dose rates. Such studies take a long time, and a long-term irradiation setup has been commissioned at LBNL, using a Kr-85 beta source. The setup is known as the SLow Irradiation of Phase-II Pixel Readout (*SLIPPER*), and the details of this measurement are described in [13]. In brief, a gaseous Kr-85 beta source with an activity of 2.22 GBq is placed on top of an RD53A chip. The achieved dose rate is estimated to be 0.025 Mrad/h and has been verified in a dedicated X-ray irradiation campaign at CERN. This dose rate corresponds to what is expected in the innermost layer of ITk during the LHC operation. The chip and source are placed inside a commercial freezer and kept cold at $-15\,°\text{C}$, as measured on the chip NTC. This setup has been continuously operating at LBNL since 2018. In the following, the results obtained with SLIPPER are compared to the X-ray irradiation results measured at Oxford.

First, the behaviour of the ring oscillators at low TID is compared between the low dose rate irradiations using X-rays and Kr-85. The SLIPPER setup uses an RD53A chip, where only a subset of the ring oscillators was available. Therefore, the SLIPPER ring oscillator results are compared to the behaviour of the Bank A ring oscillators in ITkPixV1, which utilise precisely the same gates. Figure6.16 shows the increase in ring oscillator gate delay at low TID, as measured in SLIPPER and using X-rays at a dose rate of 0.020 krad/h. A linear fit to the X-ray data is also shown to extrapolate to slightly higher TID. Good agreement can be seen between the X-ray and Kr-85 data, also when comparing numerical results of fits to both datasets.

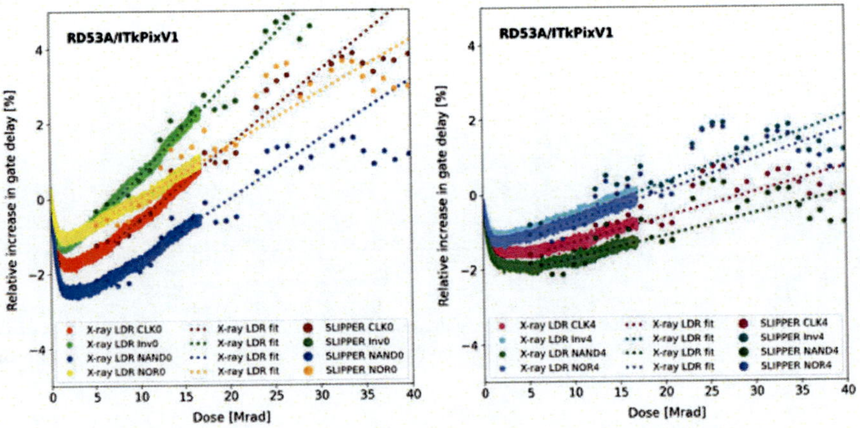

Fig. 6.16 Comparison of the measured relative increase in gate delay in X-ray and Kr-85 irradiations at low TID, for strength 0 gates (left) and strength 4 gates (right)

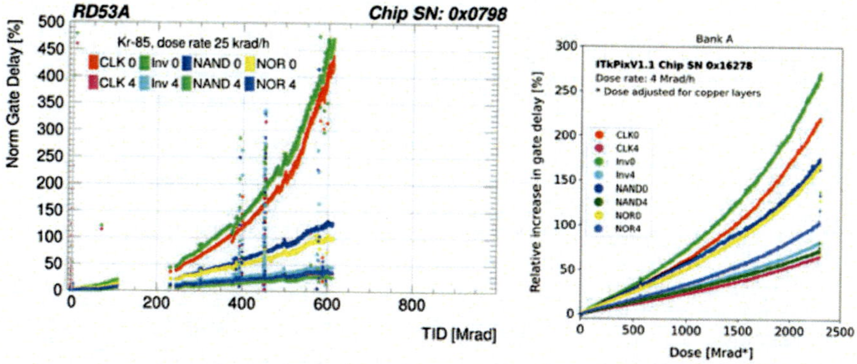

Fig. 6.17 Results from low dose rate irradiations with Kr-85 (left), from [13], and results from an X-ray irradiation campaign to 2.3 Grad at a dose rate of 4 Mrad/h (right)

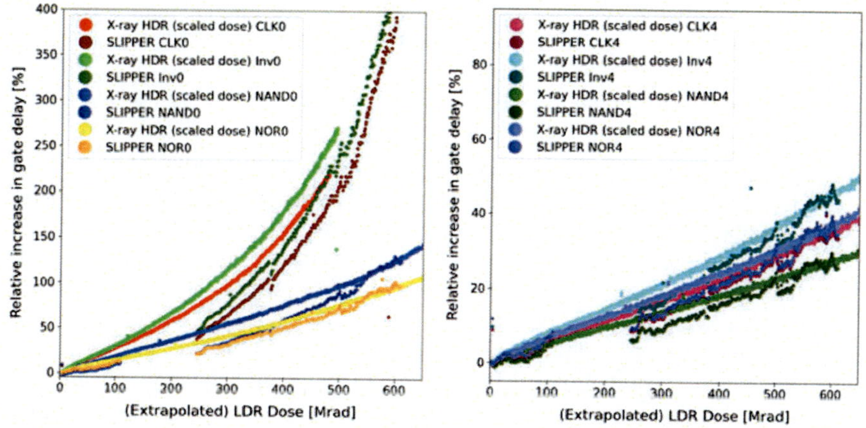

Fig. 6.18 Comparison of Kr-85 irradiations and X-ray irradiations at 4 Mrad/h for the strength 0 (left) and strength 4 (right) gates. The dose for the X-ray irradiations is scaled by the conversion factors derived in Table 6.4

Therefore, the damage observed in low dose rate X-ray irradiations is compatible with what is observed in irradiations using charged particles.

As the next step, all of the available data from the Kr-85 measurements is considered. Until now, the SLIPPER setup has collected a total of 610 Mrad of TID, and the increase in the ring oscillators gate delays is shown in Fig. 6.17 (left). It is evident that the gate delay increase is non-linear for the strength 0 gates. Such effects should be considered when evaluating the expected damage to the ITk readout chip digital logic. Similar non-linear effects can also be seen at very high doses in high dose rate X-ray irradiations. Figure 6.17 (right) shows the results of a high dose rate X-ray irradiation up to a total dose of 2.3 Grad. Here a non-linear behaviour at high doses can also be seen.

Table 6.4 Comparison of the ratio of relative gate delay increase, i.e. radiation damage, at low and high dose rates, derived using comparisons with Kr-85 and low dose rate X-ray irradiations

Ring oscillator	Delay increase ratio LDR/HDR	
	Comparison with Kr-85 at high TID	Comparison with LDR X-rays at low TID
CLK 0	4.8 ± 0.4	3.2 ± 0.2
CLK 4	2.4 ± 0.1	2.6 ± 0.3
Inv 0	4.6 ± 0.4	3.4 ± 0.2
Inv 4	2.5 ± 0.1	2.9 ± 0.2
NAND 0	3.2 ± 0.2	2.7 ± 0.3
NAND 4	1.7 ± 0.2	1.9 ± 0.3
NOR 0	2.7 ± 0.1	2.8 ± 0.2
NOR 4	1.8 ± 0.1	2.3 ± 0.2

Comparing the data from the SLIPPER setup and the X-ray irradiations, it is possible to derive another conversion factor between the damage in low and high dose rate irradiations (Fig. 6.18). This is done by considering the gate delay increases at a TID of 450 Mrad, 550 Mrad and 610 Mrad in the Kr-85 irradiation data and determining the dose with corresponding damage in the high dose rate X-ray irradiation. For example, in the case of the strength 0 NOR ring oscillators, the measured gate delay increase in SLIPPER is 108% at a TID of 550 Mrad, which corresponds to 1750 Mrad of TID in the high dose rate X-ray irradiation. Consequently, a conversion factor of 2.8 is needed between the low and high dose rate results. The average of the conversion factors for the three different doses is taken, and the results are summarised in Table 6.4. The values are also compared to the previously derived correction factors at low TID. The values agree very well for the strength 4 gates, but for the strength 0 gates, the factors derived using the SLIPPER data are significantly larger. This is likely due to the stronger non-linear behaviour of the strength 0 gates. Additionally, at very low ring oscillator frequencies, the values become more sensitive to the VDDD voltage correction and how it is applied, which might cause some discrepancies. Figure 6.19 shows the comparison of the gate delay increases obtained in the Kr-85 irradiations at low dose rate and the high dose rate X-ray irradiations, with the dose scaled by the derived correction factors. There is some difference between the shapes of the data for the strength 0 gates, but overall a reasonable agreement is seen between the two irradiation campaigns. Therefore it can be concluded that high dose rate irradiations can provide a good proxy for low dose rate irradiations, with dedicated gate-dependent correction factors applied to the high dose rate results.

As a final consideration, the expected gate delays in the ITkPixV1 digital logic can be calculated after high TID delivered at a low dose rate, corresponding to what is expected after a lifetime in HL-LHC. For this calculation, a functional form has to be determined to describe the behaviour of the gate delays at high TID, based on the values at low TID. Several functional forms were tested on the available data from the

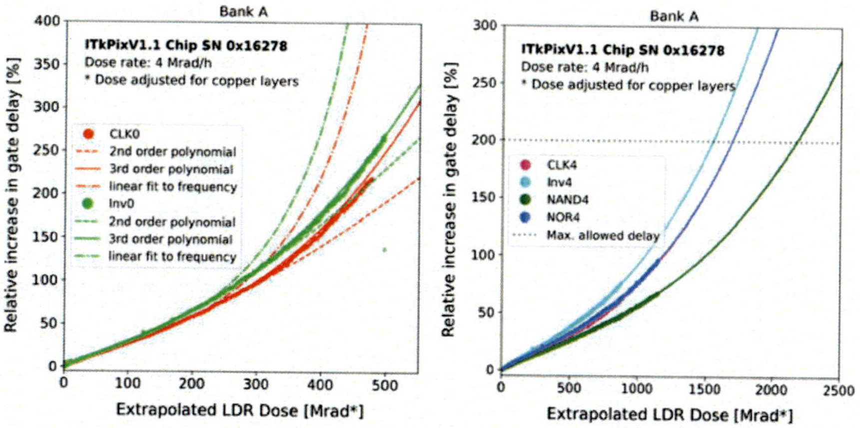

Fig. 6.19 Test of possible functions to describe the non-linear behaviour of the strength 0 ring oscillator gate delay, to extrapolate from the low dose region (<200 Mrad) to the high dose region (left). Extrapolation of the gate delays of the strength 4 gates to high total doses (right)

strength 0 gates, fitting the region below 200 Mrad of TID. A third-order polynomial was found to match the data at higher doses best, as illustrated in Fig. 6.19 (left). The expected delay of the strength 4 gates at high doses can be extrapolated using a third-order polynomial, as shown in 6.19 (right). For the nominal TID of 1 Grad after the ITk lifetime, the delays are well below the required margins of 200%. Based on this extrapolation, operational issues of the digital logic in the ITkPixV1 chip are only expected after 1.5 Grad of low dose rate irradiation in the experiment.

6.5 Radiation Damage to the Analogue Front-End

The radiation damage to the analogue front-end of the ITkPixV1 chip also needs to be understood. In particular, this relates to the behaviour of the thresholds. During detector operation, the thresholds of the readout chip will be tuned to ensure that all pixels have a similar threshold and ToT response. This is done by adjusting the global threshold (parameterised with the `Vth1` register) and the pixel threshold (adjusted with the `TDAC` parameter for each pixel). How often this will have to be done during operation depends on the radiation damage to the analogue front-end. Additionally, the behaviour of the analogue front-end after irradiation to large total doses needs to be understood. In particular, it has to be understood whether the chip is sufficiently radiation tolerant to be operational and tuneable at all expected parameter settings after irradiation.

6.5.1 Radiation Damage During Irradiation

In the first irradiation study of the analogue front-end, measurements of the chip thresholds were performed during irradiation. First, the global and pixel thresholds of the ITkPixV1 chip were tuned to 1000 e before irradiation, and the threshold distributions were continuously measured while irradiating the ITkPixV1 chip at a dose rate of 4 Mrad/h. Every 8 Mrad, the thresholds were fully tuned, and the measurements were repeated. Such a study provides insight into how quickly the threshold distribution disperses at different doses. Examples of threshold distributions during irradiation are shown in Fig. 6.20. Figure 6.20 (left) shows the threshold distributions in the first few threshold scans during the irradiation. In the beginning, the threshold distribution is strongly peaked at around the tuned value of 1000 e. With increasing dose, the threshold distributions rapidly broaden. Figure 6.20 (right) shows the same threshold distributions after retuning the chip once it reached a total dose of 8 Mrad. Of note is that here the delivered dose between the threshold scans is larger. An increase in the width of the threshold distribution is still visible, but the effect is significantly smaller than at the beginning of the irradiation.

The quantity of threshold dispersion can be defined to parametrise this effect. The dispersion is determined by considering the threshold value of each pixel and calculating how much it changes with irradiation. For each pixel threshold measurement, the pixel-by-pixel difference in threshold values is calculated with respect to the initial measurement after tuning. The distribution of these differences is then considered, and its width is defined as the *threshold dispersion*. By definition, this quantity is 0 directly after tuning the chip and is expected to increase with irradiation. Figure 6.21 (left) shows the change in threshold dispersion with irradiation. The different colours indicate the different doses at which the chip was tuned. There is a rapid increase in threshold dispersion during the first cycle, up to 200 e. The dispersion decreases for the subsequent cycles, and the behaviour stabilises after approximately 30 Mrad. This suggests that the most damage to the analogue front-end occurs during the first few Mrad of TID. In turn, this means that during the operation of HL-LHC, some

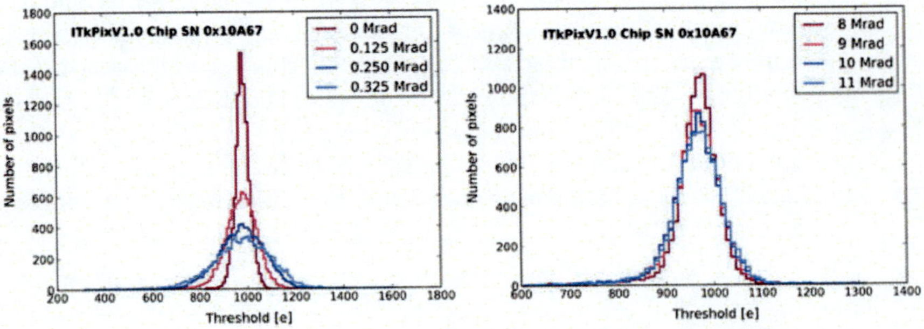

Fig. 6.20 Threshold distributions during X-ray irradiations (at a dose rate of 4 Mrad/h) after tuning the chip thresholds in the beginning of the irradiation (left) and after 8 Mrad of TID (right)

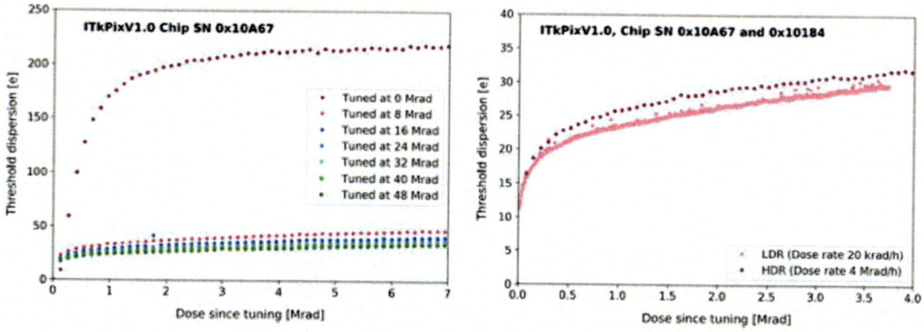

Fig. 6.21 Measurements of the threshold dispersion at a dose rate of 4 Mrad/h, after tuning the chip thresholds at different doses (left). Comparison of threshold dispersion measured at low (20 krad/h) and high (4 Mrad/h) dose rates (right)

issues with the analogue front-end are expected during the first few runs, possibly requiring more frequent retuning. After the first 30 Mrad the analogue front-end should be well-behaved, regardless of the total delivered dose.

For the digital logic, it has been observed that the radiation damage strongly depends on the dose rate. Therefore, the dose rate dependence was also studied for the analogue front-end. The expectation is that the damage to the analogue front-end should not depend strongly on dose rate, as the transistors used here are much larger. As an experimental test, threshold dispersion measurements were performed at a high dose rate of 4 Mrad/h and a low dose rate of 20 krad/h. The resulting threshold dispersion curves are shown in Fig. 6.21 (right). The threshold dispersion behaves similarly for low and high dose rates, with a rapid increase in the first Mrad after tuning, which then flattens. The overall threshold dispersion is slightly larger for the chip irradiated at a high dose rate, but this could be due to chip-to-chip differences. The increases in threshold dispersion beyond the initial rise are compatible between both measurements. This means that the behaviour of the analogue front-end after a particular delivered dose in an irradiation study at a high dose rate is representative of what is expected in HL-LHC.

6.5.2 *Radiation Damage at High TID*

Finally, the behaviour of the analogue front-end was studied at high total doses, by testing the tuned threshold distributions before and after irradiation to 1 Grad. The global and pixel thresholds are adjusted to 1000 e using the dedicated tuning algorithms. The resulting threshold and noise distributions after tuning are shown in Fig. 6.22, before irradiation and after irradiation to 1 Grad. Before irradiation, the threshold distribution has a mean of 962 e and a width of 39 e. After irradiation, the width of the distribution increases to 53 e. However, the mean of 955 e is still very close to the nominal value of the tuning. The noise distributions show only a

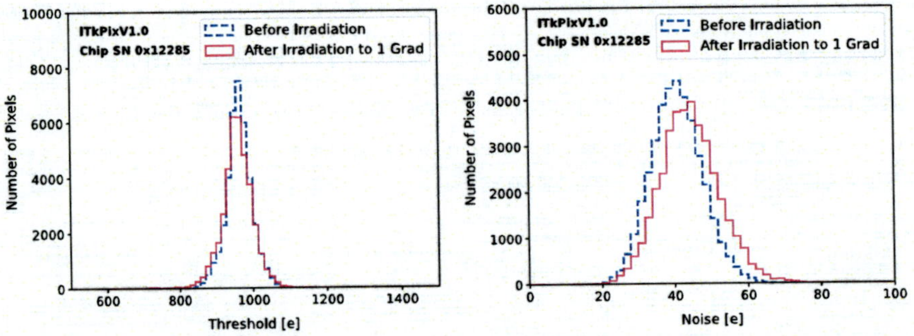

Fig. 6.22 Threshold (left) and noise (right) distributions of the ITkPixV1 chip after tuning, before and after irradiation to 1 Grad. The distribution corresponds to one quarter of the available pixels in the chip

slight increase in noise from 41 e to 44 e after irradiation. In conclusion, it is still possible to tune the thresholds of the ITkPixV1 chip well after irradiation to 1 Grad. Of note is that all of the described measurements are performed on bare chips only. The noise is expected to be larger in the detector due to the added capacitance from a sensor, as discussed in Sect. 6.7. Additional studies on the analogue front-end were performed by varying chip parameter settings and reducing the operating voltage and temperature. No issues were found down to a `VDDA` of 1.05 V and −40 °C, which are significantly more severe conditions than the expectation in ITk.

6.6 Radiation Damage to the SLDO

The next performed irradiation campaign focused on the powering of the ITkPixV1 chip. The SLDO behaviour and the evolution of the chip voltages with irradiation were studied. The irradiation was performed similarly to the previous irradiations, at a dose rate of 4 Mrad/h and a temperature of –10 °C, and the ring oscillator frequencies were monitored to validate the dose rate. In addition to this, various chip voltages and currents were measured to understand their drift with irradiation, such as the digital and analogue voltage, the pre-regulator voltages, and the reference current I_{ref}. After each 200 Mrad of the delivered dose, the chip was turned off and cooled down to around –20 °C. Then, a V-I curve was measured to characterise the start-up behaviour of the SLDO circuit at different doses and low temperatures.

Table 6.5 summarises the changes of the chip currents and voltages after irradiation to 1 Grad. The reference current I_{ref} increases by 4% throughout the irradiation. The analogue and digital currents are derived from it and therefore also increase, but there is a more considerable increase in the analogue input current than the digital one. The digital and analogue input voltages `VinA` and `VinD` only increase slightly, as expected since they should not be heavily dependent on the readout chip electronics.

Table 6.5 Increase of the currents and voltages of the ITkPixV1 chip after irradiation to 1 Grad with X-rays at a dose rate of 4 Mrad/h

Voltage/Current	Increase (%)
Currents	
Main reference current `Iref`	+4
Analogue input current `I_in analog`	+14
Digital input current `I_in digital`	+9
Analogue shunt current `I_shunt analog`	+5
Digital shunt current `I_shunt digital`	+5
Voltages	
Analogue input voltage `VinA`	+3
Digital input voltage `VinD`	+3
Analogue reference voltage `VrefA`	+12
Digital reference voltage `VrefD`	+8
Analogue LDO output voltage `VDDA`	+14
Digital LDO output voltage `VDDD`	+9
ADC reference voltage `VrefADC`	+8
SLDO offset voltage `V_offset`	+1

The most considerable drift is observed for the `VDDA` voltage, with +14 %, and the `VDDD` voltage with 9 %, which is consistent with the drifts observed for the digital and analogue currents. This behaviour is important to consider for the operation of the ITk detector, and how often these voltages will have to be adjusted in the experiment. It is also worth noting that the `V_offset` voltage stays very stable during irradiation, which can be explained by the fact that it is produced by a separate circuit and is independent of the main reference current.

Finally, the SLDO V-I start-up curves at low temperatures were measured at different total doses during irradiation. The resulting curves are shown in Fig. 6.23 for the analogue input and regulated voltages, and similar behaviour was also observed for the digital voltages. Before irradiation, the input voltage reaches its linear regime at an input current of around 1.0 A and results in a `VDDA` output voltage of approximately 1.2 V. With irradiation, this starting point of the linear SLDO regime shifts to higher input currents. This behaviour is caused by the increasing `VDDA` voltage on start-up, which is caused by the drift of the reference current with irradiation. The later start-up can be mitigated by configuring or reducing the `VDDA` later and is thus not a cause for concern for the chip operation in the experiment. Overall, even after a total dose of 1 Grad, it is still possible to power up the chip at low temperatures without any significant issues.

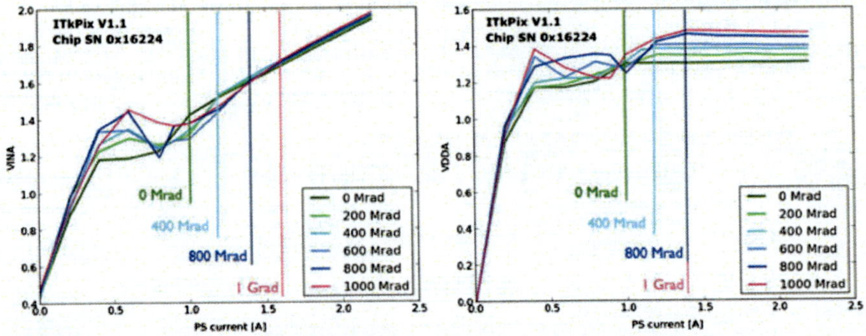

Fig. 6.23 SLDO power-up curves for the analogue voltages VinA (left) and VDDA (right). The vertical lines indicate the start of the linear SLDO regime at different TID. The expected behaviour of the SLDO circuit is shown in Fig. 6.6

6.7 Proton Irradiations of ITkPixV1 Modules

As a final test of the ITkPixV1 chip, a proton irradiation campaign of ITkPixV1 modules was performed at the Birmingham cyclotron using 27 MeV protons [14] up to a fluence of 5×10^{15} n_{eq}/cm^2. The primary goal of this irradiation campaign was to characterise the behaviour of the ITkPixV1 chips when they are assembled into modules. The chip behaviour is expected to be different when it is attached to a sensor due to the additional capacitance of the sensor, which will increase the noise. Additionally, the chip needs to be characterised for sensors with larger leakage currents after irradiation, which will further deteriorate its performance.

6.7.1 Proton Irradiation Setup

The proton irradiation campaign was performed on two ITkPixV1 modules built using 3D pixel sensors from Sintef [15] and bump-bonded at IZM [16]. The modules were then mounted on single-chip cards and characterised in detail before irradiation. The irradiation campaign was performed at the Birmingham M40 cyclotron [14]. The setup for this irradiation is shown in Fig. 6.24. The single-chip card is mounted inside a cold box, and a 1 cm Aluminium shield is placed on top of the single-chip card to avoid activation of the metal components on the card. The cold box is cooled using a Norhof liquid nitrogen evaporative cooling system. The modules were cooled down to around −30 °C before the irradiation. During the irradiation, the chip temperature increased significantly but stayed below 0 °C for the entire duration of the campaign. The chip was powered and read out while being irradiated, and the chip voltages were monitored. The chip currents were reset frequently. Although communication issues with the chip were expected during the irradiation due to the

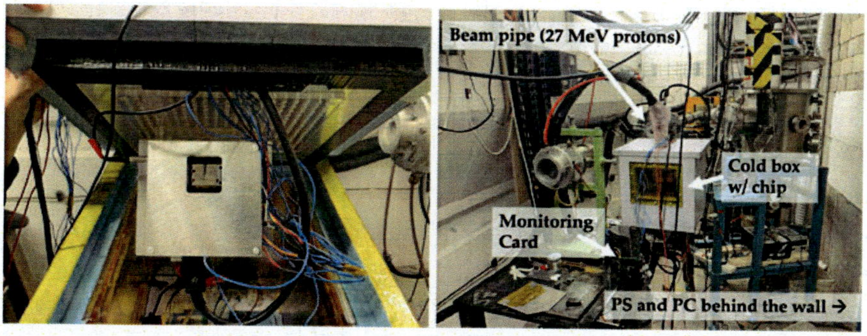

Fig. 6.24 Pictures of the setup for proton irradiation at the Bimingham Cyclotron. The ITkPixV1 modules were mounted inside a cold box, and powered and read out during the irradiation

high instantaneous proton flux, the ring oscillators were attempted to be read out throughout the irradiation. The irradiation was performed by moving the cold box through the beam to scan the 1 cm beam spot over the surface of the chip. Before irradiation, the chip was aligned with the beam, and a test run was performed to ensure good coverage of the module. During the irradiation, a Nickel foil was placed in front of the module, later used to determine the total delivered fluence. From this, the delivered fluence was determined to be 4.9 and 4.7×10^{15} n_{eq}/cm^2 for the first and second module, respectively.

6.7.2 Measurements During the Irradiation

It was possible to communicate with the chip and measure the ring oscillator frequencies throughout almost the entire irradiation, which illustrates the robustness of the ITkPixV1 chip even in an extreme environment. The resulting plot of ring oscillator frequency, VDDD and temperature is shown in Fig. 6.25 (left) for one of the modules. The temperature fluctuated significantly during the irradiation because the chip current had to be reset very frequently, and the beam spot was scanned over the surface of the module, delivering a large amount of dose in a short time. The measurement of the ring oscillator frequencies is very noisy; however, the characteristic decrease of the ring oscillator frequency is still visible.

The main advantage of reading the ring oscillators during this irradiation is that it allows to use them for dosimetry and compare them to the results obtained in X-ray irradiations. Figure 6.25 (right) shows the increase in gate delay for an X-ray irradiation campaign at a dose rate of 4 Mrad/h in black. The ring oscillator delay data during the proton irradiation is also plotted, with the dose rate parameter adjusted such that data matches the X-ray results as well as possible. Using these results, it is now possible to determine the expected TID by comparing the endpoint of the

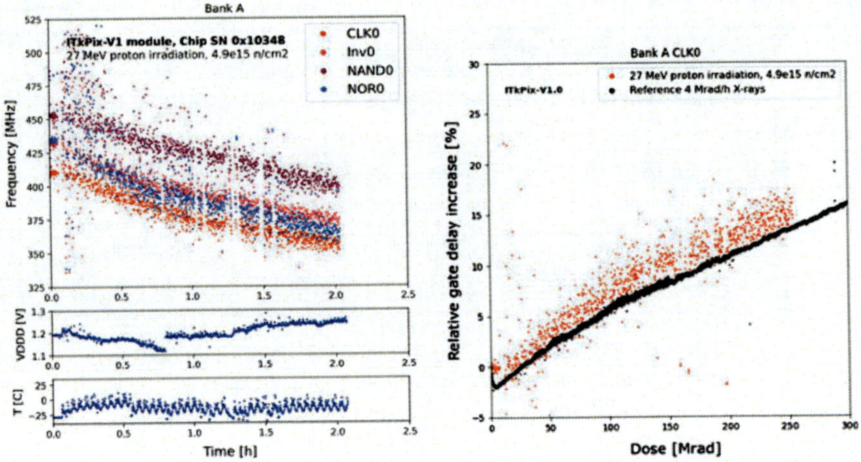

Fig. 6.25 Ring oscillator frequencies for the strength 0 gates during proton irradiation of ITkPixV1 module 0x10348 (left). Ring oscillator delay increases during proton irradiation compared to an X-ray irradiation at a dose rate of 4 Mrad/h for the CLK 0 ring oscillators (right)

proton irradiation data to the X-ray irradiation data. Based on this estimate, the first and second modules received a TID of around 250 and 220 Mrad, respectively.

It is also possible to estimate the TID delivered in the proton irradiation by converting the measured fluence into TID using the Bethe model for protons [17]. Following this approach, the expected TID is approximately 500 Mrad, twice as large as the value estimated from the ring oscillator measurements. There is no clear explanation for this, but there can be several possible reasons. Most likely, this discrepancy is caused by the impact of having vastly different dose rates in the X-ray and proton irradiations, since the dose rate in the proton irradiation would be about 400 Mrad/h. However, this is very difficult to verify, as X-ray setups with high dose rates are unavailable. A further explanation could also include the temperature dependence of the radiation damage, as the temperature in the proton irradiation varied significantly. However, the exact TID value achieved in the proton irradiation campaign is not critical. The main focus of the measurement is the analogue front-end, where most of the radiation damage occurs at the beginning of the irradiation. Additionally, the bulk damage to the sensor is of interest, and the total fluence has been precisely measured during the irradiation.

6.7.3 *Measurements of the Analogue Front-End After Irradiation*

After the proton irradiation, the modules were used to characterise the behaviour of the analogue front-end when attached to a sensor with a significant amount of leakage current. Two configurations of leakage current were tested: 2 nA and 10 nA per pixel, representing two different points in the lifetime of the ITk detector, where 2 nA is a realistic operational scenario, and 10 nA is a worst-case scenario. As a first step, the behaviour of the sensors after irradiation was characterised. The currents for different bias voltages (i.e. the IV curves) for the two modules are shown in Fig. 6.26, for different temperatures. The leakage currents are of the expected order of several hundred μA, and show the characteristic increase with temperature. The characterisation of the analogue front-end was performed for both module 0x10348 and 0x10358, and the results were found to be consistent, so for simplicity, only results for module 0x10348 are shown going forward.

With these particular modules, further considerations have to be made when characterising them. Due to issues with the bump bonding, only a subset of the pixels are connected, as can be seen in the source scan in Fig. 6.27 (left). Furthermore, during the irradiation the beam and the devices were not perfectly aligned, so the irradiation missed the top region of the chip, as can be seen in the discoloured area in Fig. 6.27 (center). As a consequence, the module can be divided into different areas: *connected + irradiated*, *connected + unirradiated*, *disconnected + irradiated*, and *disconnected + unirradiated* pixels. For most of the results, only the *connected + irradiated* pixels are considered, and the leakage current per pixel is calculated only taking these pixels into account. For the measurements at the different leakage current settings of 2 nA and 10 nA per pixel, the leakage current is adjusted by varying the bias voltage. The measurements are taken at various temperatures and the bias voltages are adjusted accordingly to keep the leakage current constant.

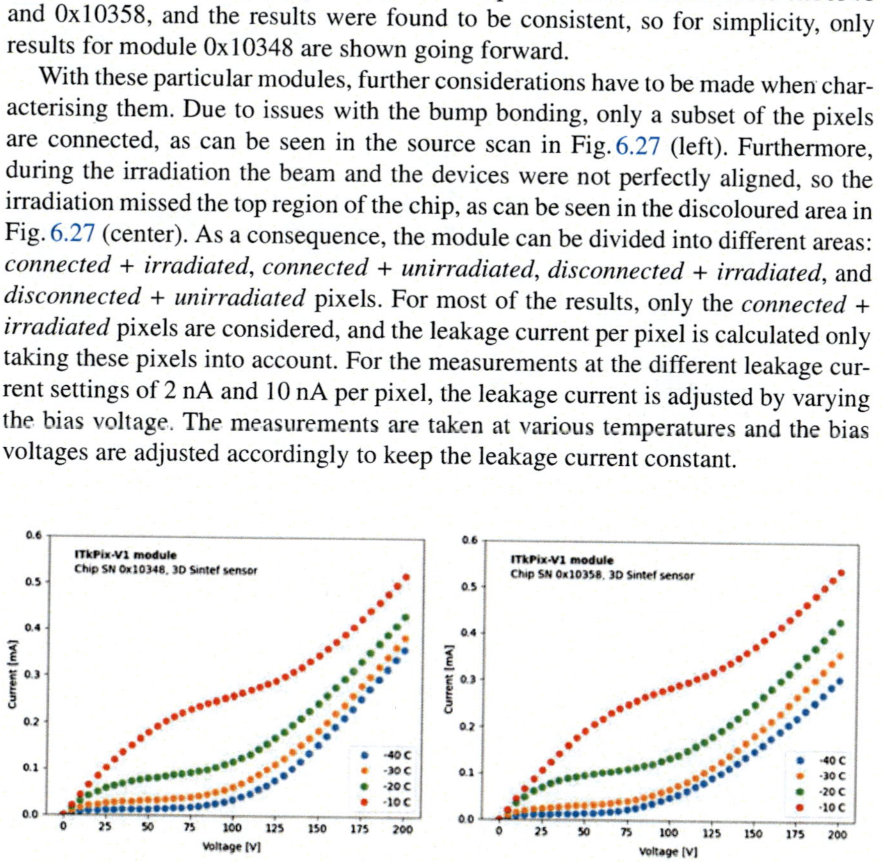

Fig. 6.26 IV curves after irradiation at different temperatures for module 0x10348 (left) and 0x10358 (right)

Fig. 6.27 Illustrations of the areas of the irradiated module. A source scan for module 0x10348 shows the connected and disconnected pixels (left). A picture of the irradiated single-chip card where discolouration illustrates the irradiated area (center). An illustration of the different areas considered areas in the module (right)

First, measurements were performed using analog scans and threshold scans without the tuning of the pixel thresholds. These results are not discussed in detail in this thesis. In summary, no significant issues of chip operation were found after irradiations, and the analogue front-end was fully functioning down to −40 °C and with a reduced analogue voltage of 1.07 V. As a final validation of chip performance, the tuning of the pixel thresholds was tested at the different leakage currents, without and with the leakage current compensation circuit (*LCC*), and the results are summarised in the following.

Measurements after Tuning of Pixel Thresholds at Leakage Currents of 2 nA/pixel

Due to the different areas of the modules in terms of connected bumps and irradiation, and the relatively high noise, the tuning was performed in several steps:

- The global threshold was set to 2000 *e* by hand based on the threshold distribution of the *connected + irradiated* pixels only.
- The pixel threshold was tuned to 2000 *e* using the standard pixel threshold tuning algorithm.
- The global threshold was set to 1500 *e* by hand, based on the *connected + irradiated* pixels only.
- The pixel threshold was tuned to 1500 *e*.
- A threshold scan and an analogue scan were performed.

Figure 6.28 shows two examples of ToT maps from analogue scans after this tuning procedure, without (left) and with the LCC switched on (right). For both scans, most pixels are responsive, but noticeable differences can be seen between the connected and disconnected and the irradiated and unirradiated pixels. In particular, the mean ToT is higher for the irradiated pixels and highest for the irradiated and

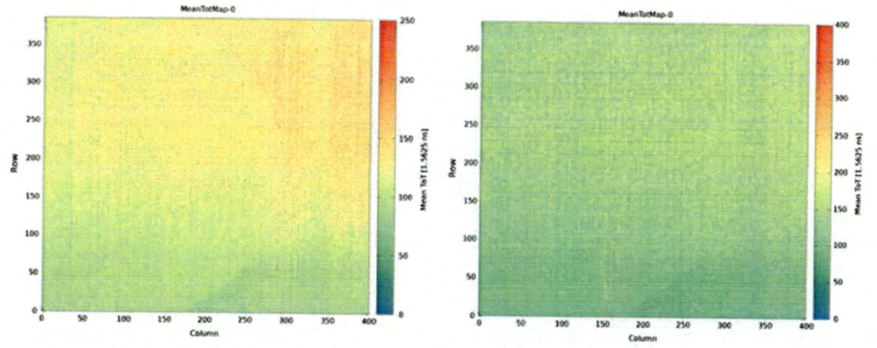

Fig. 6.28 Maps of the ToT measured in analogue scans of a proton-irradiated module, after tuning the thresholds at a leakage current of 2 nA/pixel, with the LCC off (left) and LCC on (right)

disconnected pixels. The difference in the behaviour of the pixels is reduced with the LCC switched on.

An example of the threshold scan results for the different areas of the module, for a leakage current of 2 nA/pixel at a temperature of −5 °C, can be found in Fig. 6.29. Here, both the two-dimensional maps of the quantities, as well as the distributions are shown. First, the threshold map and distribution show the expected threshold of 1500 *e* and good agreement between all four areas of the module. In the noise map, a clear distinction between the four areas of the chip can be found. The *connected+unirradiated* pixels have a larger noise of around 100 *e* compared to the *connected + irradiated* with around 80 *e*. This feature can be explained by the fact that the tuning is performed with only the *connected + irradiated* pixels. The disconnected pixels have a significantly lower noise of around 50 *e*, as expected.

Finally, these studies can be used to identify the operating range of the modules by reducing the temperature. When lowering the temperature, no operational issues were observed down to −30 °C. Figure 6.30 shows the threshold tuning results for the *connected + irradiated* pixels for 2 nA/pixel for a temperature of −40°C, for different LCC settings. At −40 °C, issues start to occur in the tuning without the LCC—the threshold distribution becomes wider, and the noise is higher. This observation is consistent with the previously described analogue scan measurements, where operational issues were observed at −40 °C and low voltages. However, the chip behaviour can be improved by switching on the LCC, as shown in the purple curve, where it is possible to recover the expected threshold and noise distribution.

Measurements after Tuning of Pixel Thresholds at Leakage Currents of 10 nA/pixel

A similar set of measurements was taken for a leakage current of 10 nA/pixel after tuning the pixel thresholds for each configuration. The ToT maps resulting from

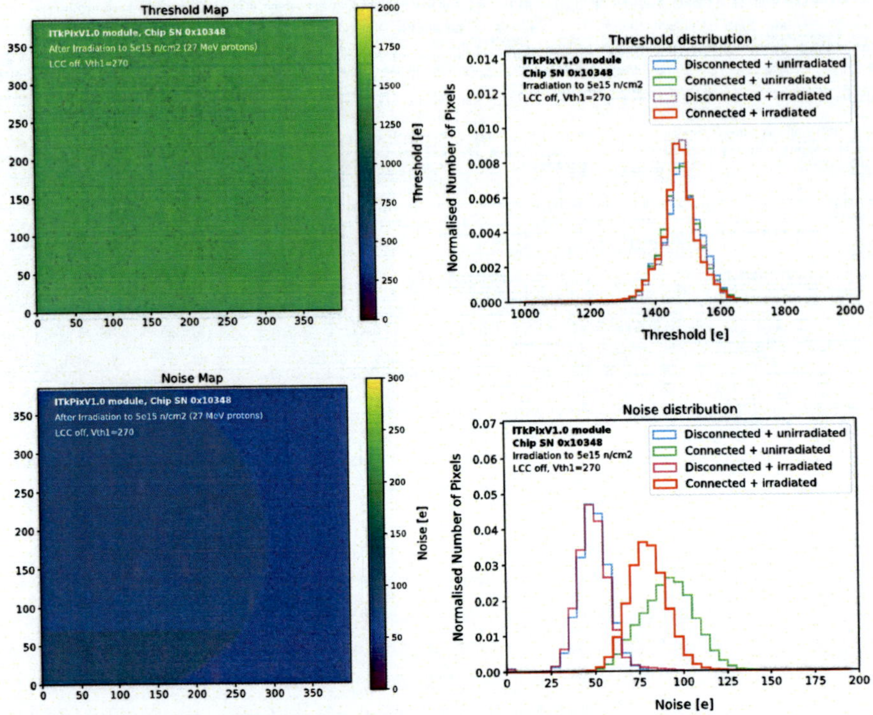

Fig. 6.29 Threshold (top) and noise (bottom) maps and binned distributions as measured on the proton-irradiated module 0x10348, after tuning the thresholds to a value of 1500 *e*. The leakage current is set to be 2 nA/pixel. Binned distributions are shown separately for the different areas in the module

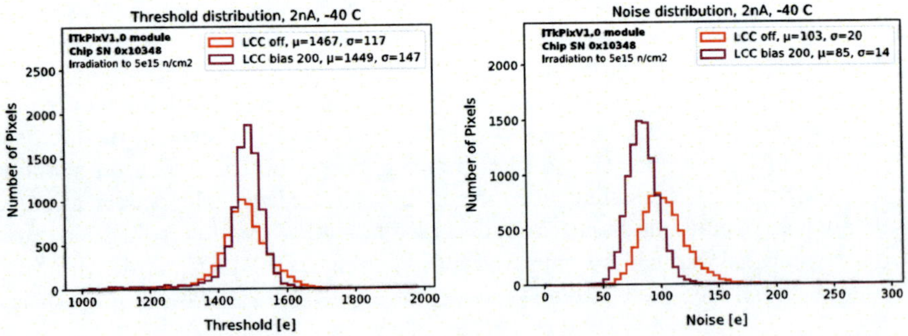

Fig. 6.30 Threshold (left) and noise (right) distributions for the connected and irradiated pixels in module 0x10348, after tuning the thresholds to a value of 1500 *e*, with and without the LCC. The leakage current is set to be 2 nA/pixel and the temperature is −40 °C

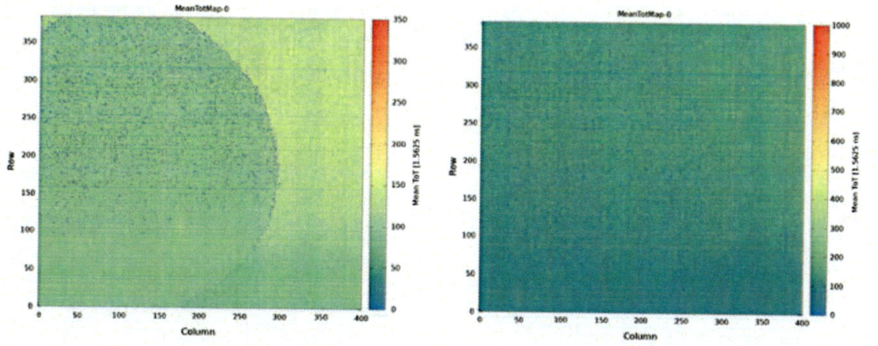

Fig. 6.31 Maps of the ToT measured in analogue scans of a proton-irradiated module, after tuning the thresholds at a leakage current of 10 nA/pixel, with the LCC off (left) and LCC on (right)

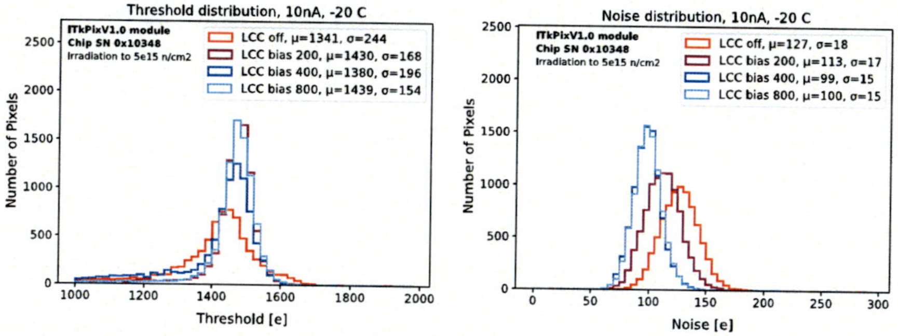

Fig. 6.32 Threshold (left) and noise (right) distributions for the connected and irradiated pixels in module 0x10348, after tuning to the thresholds to a value of 1500 e, with and without the LCC. The leakage current is set to be 10 nA/pixel and the temperature is $-20\,°\mathrm{C}$

analogue scans with and without the LCC are shown in Fig. 6.31. Here a significant number of failing pixels can be seen for the *connected + irradiated* pixels when the LCC is switched off. This feature significantly improves after switching on the LCC. Once again, there is a significant difference between the different areas of the chip in terms of observed ToT values—for the disconnected pixels, the measured ToT values are higher. With the LCC switched on, this feature is reduced, and the ToT map again looks more similar for the connected and disconnected pixels.

The threshold results, when tuning with a leakage current of 10 nA/pixel, are shown in Fig. 6.32 for the *connected + irradiated* pixels only, at a temperature of $-20\,°\mathrm{C}$. The threshold distribution is wider than the target distribution and the noise is high while the LCC is off. Once the LCC is switched on, the nominal chip performance can be recovered, and a high LCC setting gives the best results, as expected from simulation. Similar behaviour is also observed at a temperature of $-5\,°\mathrm{C}$.

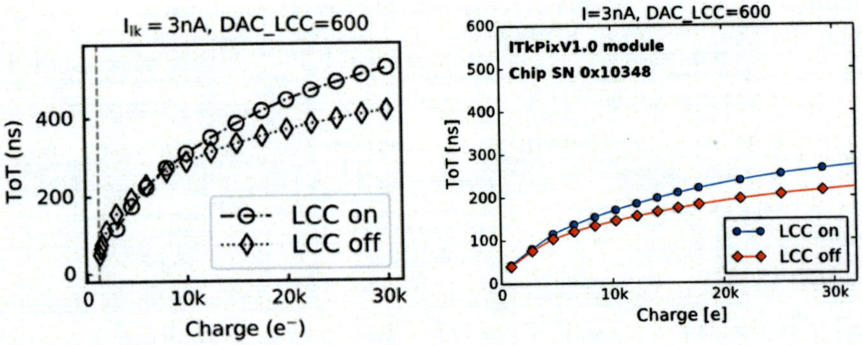

Fig. 6.33 Simulated (left) and measured (right) ToT as a function of injected charge for a leakage current of 3 nA/pixel. The simulated graph is from [2]. A typical signal in a 150 μm thick sensor is of the order of 10000 e, and is usually split over several pixels

Impact of the Leakage Current Compensation in Comparison with Simulation

Finally, the LCC circuit was characterised in more detail by studying the ToT as a function of injected charge, with and without the LCC switched on. This study has been performed in simulation for a leakage current of 3 nA/pixel, with the resulting curves shown in Fig. 6.33. Without the LCC, the response to increasing charges flattens at high total charges. In contrast, there is a more significant increase with increasing injected charge with the LCC switched on. Using the proton-irradiated modules, similar curves were measured, using the same leakage current (3 nA/pixel) and LCC setting (LCC bias 600) as in simulation. The resulting curves are shown on the right in Fig. 6.33. The absolute values of ToT are slightly different, likely due to slightly different settings on the chip feedback current. Nevertheless, a similar effect of switching on the LCC can be reproduced.

6.8 Summary and Outlook

This chapter outlined investigations into the radiation tolerance of the ITkPixV1 readout chips, which are currently the latest prototype chips for the ATLAS ITk pixel upgrade.

Several irradiation campaigns were performed using the X-ray irradiation setup at Oxford. The setup was commissioned and calibrated as part of this project. In particular, the dose rate and beam spot were calibrated using a diode and Medipix detector. The setup was found to have the ability to deliver dose rares of around 4 Mrad/h for an 8 cm beam spot with high uniformity.

The TID damage to the digital logic of the ITkPixV1 chips was studied by measuring the increase in gate delay of the ring oscillator test circuits on the chip. A total of 15 irradiation campaigns was performed using ITkPixV1 chips.

At high dose rates, an increase in ring oscillator gate delay of up to 40% was found at total doses of 1 Grad (for the gates representative of the digital logic on the chip). The dose rate dependence of the radiation damage was studied using X-ray irradiations at low dose rates and low TID, and a comparison with a long-term low dose rate irradiation campaign using Kr-85. For the ring oscillators representing the digital logic in the chip, a factor up to 2.5 higher damage was found at low dose rates. Low dose rate effects should be studied further, especially at high TID. However, for the ITk upgrade, the results are well understood and low dose rate effects are expected to increase the radiation damage within the allowed margins.

The analog front-end was also characterised in X-ray irradiations. During irradiation, the threshold dispersion was measured, and most damage to the analog front-end occured in the first tens of Mrad of TID. Furthermore, no significant dose rate effects were found in the analog front-end since larger transistors are used in this component. After irradiation to 1 Grad, the analog front-end was characterised and found to be fully operational, even down to reduced analog voltages of 1.05 V.

The powering circuit of the ITkPixV1 chips (SLDO) was characterised in a dedicated study. During irradiation to 1 Grad, the chip reference voltages increased by up to 15%. The SLDO circuit remained fully functional. However, the regular operational setting is reached only at a higher supplied current due to the voltage drift, which will have to be considered during detector operation.

Finally, the ITkPixV1 chip was characterised in modules with attached sensors in a proton irradiation campaign at the Birmingham cyclotron. Two modules were irradiated to a total fluence of $5\times10^{15}\mathrm{n_{eq}/cm^2}$. Communication with the chip was possible throughout the irradiation, and the ring oscillators and chip voltages were monitored. The comparison of ring oscillator data to X-ray irradiations suggests that the modules received a TID of around 250 Mrad. The analog front-end of the chip was characterised after irradiation for leakage currents of 2 nA and 10 nA/pixel. For 2 nA/pixel the chip is operational down to $-30\,^{\circ}$C, even with a reduced voltage of 1.1V. For 10 nA/pixel the chip is also fully functional when switching on the leakage current compensation circuit.

In summary, the ITkPixV1 chip is as radiation tolerant as expected and should survive the intended lifetime in the ITk upgrade. As the next steps of the readout chip development for the ITk upgrade, the final production version of the chip, ITkPixV2, will be submitted in 2022. The final chip will then be characterised in detail, particularly in larger-scale system tests and beam test measurements, to understand the device in more realistic operational scenarios.

References

1. Garcia-Sciveres M (2017) The RD53A integrated circuit. Tech rep CERN-RD53-PUB-17-001. CERN, Geneva. https://cds.cern.ch/record/2287593
2. Garcia-Sciveres M, Loddo F, Christiansen J (2019) RD53B manual. Tech rep CERN-RD53-PUB-19-002. CERN, Geneva. https://cds.cern.ch/record/2665301
3. Equipment supplies, X-ray system, computed radiography, magnaflux (2020). https://www.euroteck.co.uk/
4. Deslattes RD et al (2005) X-ray transition energies (version 1.2). National Institute of Standards and Technology, Gaithersburg, MD
5. Koch H, Faccio F, Borghello G (2018) CERN PIN diode calibration document. https://indico.cernch/event/956616/contributions/4068849/attachments/2126064/3579530/PIN_diode_calibration_2018.pdf
6. Plackett R et al (2013) Merlin: a fast versatile readout system for Medipix3. J Instrum 8(01):C01038–C01038. https://doi.org/10.1088/1748-0221/8/01/c01038
7. Karagounis M et al (2009) An integrated Shunt-LDO regulator for serial powered systems. In: 2009 proceedings of ESSCIRC. pp 276–279. https://doi.org/10.1109/ESSCIRC.2009.5325974
8. Heim T (2017) YARR—A PCIe based readout concept for current and future ATLAS pixel modules. J Phys Conf Ser 898:032053. https://doi.org/10.1088/1742-6596/898/3/032053
9. Faccio F et al (2015) Radiation-induced short channel (RISCE) and narrow channel (RINCE) effects in 65 and 130 nm MOSFETs. IEEE Trans Nucl Sci 62(6):2933–2940. https://doi.org/10.1109/TNS.2015.2492778
10. Huber CC-508 chiller. https://www.huber-online.com/download/pdf_datasheets/englisch/2045.0002.01.PDF
11. NTC datasheet (2021). https://product.tdk.com/en/search/sensor/ntc/chip-ntc-thermistor/info?part_no=NTCG103JF103FTDS
12. Borghello G et al (2018) Dose-rate sensitivity of 65-nm MOSFETs exposed to ultrahigh Doses. IEEE Trans Nucl Sci 65(8):1482–1487. https://doi.org/10.1109/TNS.2018.2828142
13. Dimitrevska A (2021) Low dose rate irradiation of the RD53A chip with Kr-85 beta source. In: TWEPP 2021, topical workshop on electronics for particle physics
14. Allport P et al (2017) Recent results and experience with the Birmingham MC40 irradiation facility. JINST 12(03). https://doi.org/10.1088/1748-0221/12/03/c03075
15. SINTEF MiNaLab, P.O. Box 124, Blindern, 0314 Oslo, Norway
16. Fraunhofer IZM, Gustav-Meyer-Allee 25, Gesundbrunnen, 13355 Berlin, Germany
17. Displacement damage in silicon. https://rd50.web.cern.ch/NIEL/

Chapter 7
X-Ray Measurements of the Monolithic mini-MALTA Prototype

The final chapter of this thesis focuses on potential novel pixel detector technologies for the ATLAS detector and future collider applications. There is a clear need to futher upgrade our detector performance to improve our sensitivity for physics measurements, such as the $VH(H \rightarrow c\bar{c})$ search described in Chap. 4. In particular, the $VH(H \rightarrow c\bar{c})$ search would benefit from better resolution of the inner tracker, which would help with momentum and vertex reconstruction. Such an improvement would, in turn, enhance the performance of our flavour tagging algorithms and thus the identification of heavy-flavour jets in the detector. The performance of a pixel detector in a collider experiment can be improved by reducing the pixel pitch and the detector material. A reduced pixel pitch will result in a better position resolution and, therefore, better resolution in the reconstruction of impact parameters. A material reduction will result in less multiple scattering, improving the position and momentum resolution, where the former is a main ingredient in vertex reconstruction.

Monolithic detectors provide a potential option to achieve these goals. In monolithic pixel detectors, the active area and readout electronics are located in the same piece of silicon. This makes it possible to reduce the thickness of the device significantly, as no additional readout chip and bump bonds are needed. Removing the bump bonding from the production process reduces the cost significantly and allows for smaller pixels. These advantages make monolithic detectors an excellent candidate for future pixel detectors. On the downside, the technology is not as well developed as hybrid pixel detectors and, in general, not as radiation tolerant. At the LHC, monolithic detectors have been first developed within the ALICE collaboration [1]. Based on this technology, monolithic detectors have also been considered for the outermost layers of the ITk pixel detector upgrade, and this chapter summarises some work performed within this effort. In particular, this chapter discusses the measurements of the pixel performance of the monolithic mini-MALTA prototype in an X-ray beam test at Diamond Light Source. The pixel response is measured with high precision, and the performance of the different pixel designs within the prototype is compared.

M. Mironova, *Search for Higgs Boson Decays to Charm Quarks with the ATLAS Experiment and Development of Novel Silicon Pixel Detectors*, Springer Theses,
https://doi.org/10.1007/978-3-031-36220-0_7

Additionally, measurements of the pixel response at varying levels of irradiation are described, up to the total doses that would be expected in the outermost layer of ITk (1×10^{15} n_{eq}/cm^2).

7.1 The mini-MALTA Prototype

The mini-MALTA device is a depleted monolithic pixel sensor prototype made in TowerJazz 180 nm CIS (CMOS Image Sensor) technology, with 64×16 pixels and 36.4×36.4 μm^2 pixel pitch. More details about the device, laboratory and particle beam test measurements can be found in [2]. It is based on the technology of the ALPIDE chip, developed for the ALICE ITS upgrade [3]. The ALPIDE design was then adapted to allow for full depletion of the epitaxial layer [1]. This was tested in the MALTA chip, a full-sized CMOS demonstrator for the ATLAS ITk upgrade [4, 5]. Mini-MALTA was designed as a test chip based on the MALTA chip to test different front-end designs, reset mechanisms and processes. Three possible approaches are tested in the mini-MALTA prototype, which are illustrated in Fig. 7.1. The standard design, similar to the one of the MALTA chip, is referred to as the *continuous n-layer*. Additionally, two modified layouts have been introduced in mini-MALTA. These process modifications are included in each pixel to improve charge collection at the pixel edges. The first variant has an *extra deep implant of p-type silicon*, which is expected to shape the electric field within the detector such that the produced charge carriers are steered more directly towards the collection electrode in the centre of the pixel. The second layout has a *gap in the n-layer*, which is expected to have a similar effect on the electric field. Both devices have been shown to perform well in TCAD simulations [6]. Each configuration is implemented with a standard front-end and a modified front-end featuring enlarged transistors to reduce random telegraph signal (RTS) noise. For the continuous n-layer configuration, the sector with the standard front-end was tested, as this corresponds to what was previously implemented in

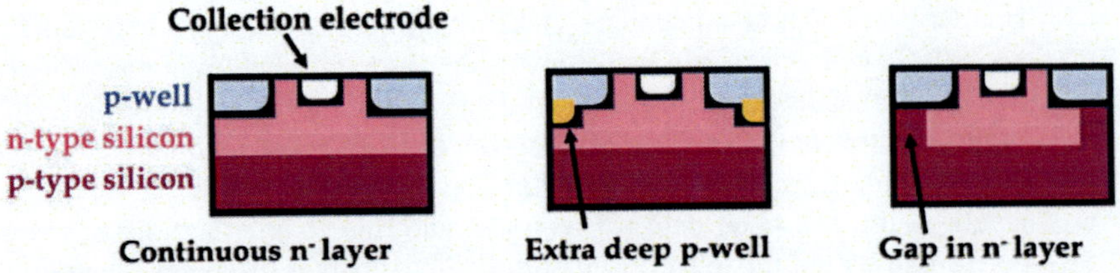

Fig. 7.1 Standard MALTA design and process modifications in mini-MALTA. The figure is not to scale, the thin n-type layer located on top of a p-type epitaxial layer with a depth of 30 μm. This layer is expected to be fully depleted, and is grown on a highly doped p-substrate which makes no contribution to the charge collection

Table 7.1 Table of mini-MALTA chips with their respective delivered dose in neutron equivalent fluence and TID

Sample	Epitaxial thickness (μm)	Irradiation	Fluence (n_{eq}/cm^2)	TID (MRad)	Annealing
W2R11	30	None	0	0	None
W2R9	30	23 MeV protons	5.0×10^{14}	66	None
W2R1	30	1 MeV neutrons	1×10^{15}		None
W5R9	25	23 MeV protons	5.6×10^{14}	74	120 m at 35 °C
W4R9	25	23 MeV protons	7×10^{13}	9	None

the MALTA chip. For the n-gap and extra deep p-well designs, the sectors with the enlarged transistors were measured, as these have been shown to perform better.

The main goal of the beam test at Diamond Light Source [7] was to study the performance of the two new layouts compared to the standard layout for different levels of irradiation. One unirradiated chip was tested (W2R11). Several irradiated devices were tested, as summarised in Table 7.1. One sample (W2R1) was irradiated to a total fluence of 1×10^{15} n_{eq}/cm^2 using neutrons from the TRIGA research reactor in Ljubliana [8]. The remaining three chips were irradiated with 23 MeV protons at the MC40 cyclotron at Birmingham to 5×10^{14} and 7×10^{13} n_{eq}/cm^2 [9]. Here the delivered fluence was estimated through nickel dosimetry, and the TID was calculated using the Bethe model for 23 MeV protons [10]. After irradiation, the behaviour of these devices was characterised by noise scans, threshold scans and data acquisition with a Sr-90 source. For one of the samples (W5R9), the noise was too high after irradiation, and its performance was improved through annealing at 35 °C for two hours.

7.2 Experimental Setup

The X-ray beam test was performed at the B16 beamline at Diamond Light Source. The experimental setup in the beamline is shown in Fig. 7.2 (left). The mini-MALTA chip was mounted on a PCB and attached to the cooling system. The cooling system was developed especially for this beam test and is shown in Fig. 7.2 (right). It consists of a water-cooled metal chuck reaching a temperature of 10 °C, coupled to the warm side of a Peltier element connected to a cold adapter plate onto which the mini-MALTA PCB was attached. This cooling system consistently kept the sensor at −13 °C, and the temperature on the cooling chuck was monitored with a temperature sensor. The setup was enclosed in a foam box, which was flushed with nitrogen to avoid condensation.

The X-ray beam was focused using a Compound Reflective Lens (CRL) X-ray mirror arrangement to a beam spot of 2 mm (FWHM), as measured using a knife-

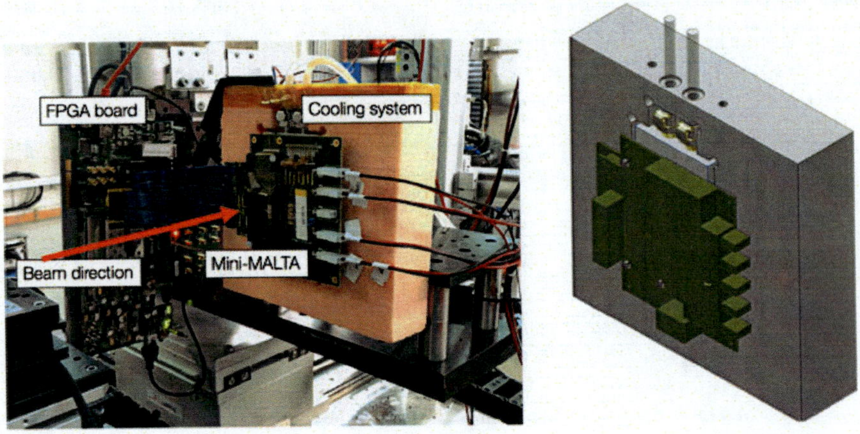

Fig. 7.2 Experimental Setup in the beamline (left). The mini-MALTA chip is cooled and located on a PCB, which is mounted onto a motion stage. CAD drawing of the cooling system with the attached mini-MALTA PCB (right)

edge technique. The mini-MALTA chip was mounted on a motion stage, which was placed at the focal point of the X-ray beam. The device was read out using a Kintex KC705 FPGA board.

The choice of beam energy was made based on an estimate of how much energy a minimum ionising particle (MIP) would deposit in the depletion region of mini-MALTA. Mini-MALTA is expected to have a depletion depth of between 20 and 25 mm, depending on several silicon properties, the bias voltage, and the irradiation levels. The given depletion depth means that a MIP would deposit between 7.7 and 9.7 keV in the device, according to Eq. 5.2. At these energies, X-rays interact primarily through the photoelectric effect and deposit all their energy in one location. Therefore, a beam energy of 8 keV was chosen for this beam test.

The beam was attenuated by 0.5 mm of aluminium to decrease the data rates of the chip. During the measurements, the detector was moved in both orthogonal directions to the beam in 2 μm steps. The precision of the movement of the motion stages is 400 nm. Data was taken for 1 s at each point to reach sufficiently high statistics. The scans covered a total area of $100 \times 100\ \mu\text{m}^2$, which fully contained at least four pixels.

The chip threshold for each measurement was chosen depending on the level of noise in the chip. The threshold values were set with the discriminator bias current (*IDB*). Then, the threshold distributions were measured with a threshold scan for each of the sectors of each device. The results for the different chips and sectors are summarised in Table 7.2. In general, larger thresholds are needed for devices that received TID damage. The precise value of the chip threshold does not impact the results of the X-ray beam test because the photon always deposits all of its energy

Table 7.2 Measured thresholds for the different samples and sectors

Sample	Fluence	n_{eq}/cm^2 TID (MRad)	Threshold [e]		
			Continuous n-layer	Extra deep p-well	n-gap
W2R11	0	0	368	197	190
W2R9	5×10^{14} (p)	66	533	303	274
W2R1	1×10^{15} (n)		330	181	187
W5R9	5×10^{14} (p)	70	681	379	374
W4R9	7×10^{13} (p)	9	437	234	232

when it interacts. For an 8 keV photon, one would expect 2000 e to be produced, significantly higher than the thresholds noted in Table 7.2, even if the charge is split between multiple pixels.

7.3 Results

7.3.1 Analysis and Single Pixel Response

The number of hits in a particular pixel was stored at each stage position during the beam test. This information was then translated into a two-dimensional map, with an example of this shown in Fig. 7.3 (left). The shape of the response of an individual pixel is visible. Next, the centre of each pixel was identified by applying a Gaussian filter [11] to the image and finding the location with the maximum value. Then, the average number of hits was calculated in a small region around the pixel centre, which is outlined with a green square. The number of hits in each pixel was then normalised to that average. This means that the response of the individual pixel is normalised individually, which removes pixel-to-pixel variations in response. Subsequently, the average normalised number of hits was calculated within a 36.4×36.4 mm^2 square, which is shown in red in Fig. 7.3 (left) and corresponds to the nominal size of the pixel. This average is defined as the relative *photon pixel response*. The uncertainty on the pixel response is defined as the standard deviation of the number of hits in the normalisation area, i.e. the green square. The photon pixel response measures how well the device responds to particles. However, it is different from the standard measure of efficiency quoted for beam tests with charged particles due to the different energy deposition for photons and charged particles and because there is no measurement of the total number of photons entering the device. A more detailed discussion of the difference between photon and charged particle beam tests can be found in [12].

The pixel response calculation was repeated for each pixel, and the pixel responses were averaged across the fully visible pixels in the scan. The single-pixel maps

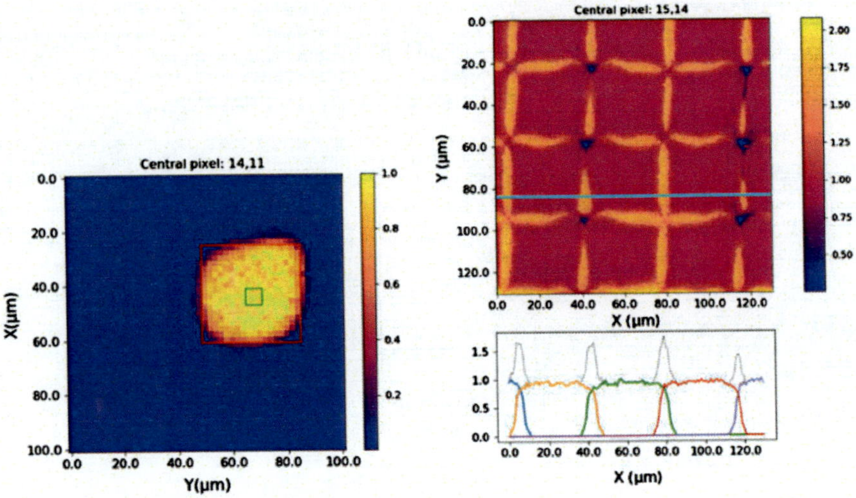

Fig. 7.3 llustration of technique used for the analysis for a single pixel of the continuous n-layer on W2R11 (left). The green square outlines the centre of the pixel, which the number of hits is normalised to. The red square outlines the expected pixel area. Example of a summed pixel response map and profile of the response along the blue line for the extra deep p-well sector of W2R1 (right). The response for each pixel is shown in a different color and the total normalised response is shown as a dashed line

are then added together while removing all entries from the pixel maps below 7% normalised response, as those hits are likely to stem from the X-ray halo or noise in the detector. An example of the summed pixel response map is shown in Fig. 7.3 (right). At the pixel edges, the produced charge can be picked up by two pixels simultaneously, which leads to a pixel response larger than 1.

An interesting feature of the mini-MALTA is that the pixel response is asymmetric and extended in one direction. A double-column structure is also visible, i.e. two adjacent columns of pixels appear mirrored. These features agree with observations in previous beam tests of the MALTA chip [13]. The asymmetric pixel shape is caused by the shape of the deep p-well cut-out. This is the region between the collection electrode and the p-well which contains n-type silicon. The shape of the deep p-well cut-out is shown as an overlay in Fig. 7.4, with the n-type collection electrode as the pink dot in the centre. In regions with less p-well coverage, i.e. a larger gap, one can observe a higher pixel response. A larger p-well gap causes a more significant potential difference with respect to the collection electrode. The p-well layout is mirrored in double columns explaining the double-column structure of the response.

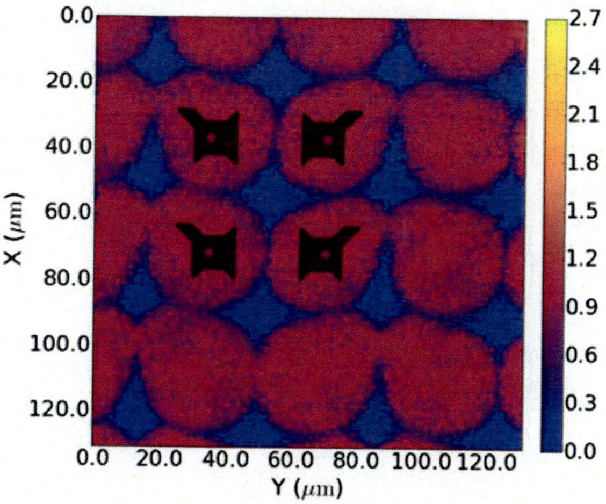

Fig. 7.4 Overlay of the scan of the continuous n-layer sector in W2R1 (irradiated to 1×10^{15} n_{eq}/cm^2) and the shape of the deep p-well cutout and collection electrodes. The cutout of the deep p-well influences the charge collection shape

7.3.2 *Pixel Response as a Function of Radiation*

The results of the normalised pixel response analysis for the different samples are shown in Table 7.3 and the examples of response maps are shown in Fig. 7.5a–i. All measurements were performed at a bias voltage of –5 V. The most accurate comparison between different designs and irradiation levels is achieved by comparing the samples W2R11, W2R9, and W2R1. These samples are from the same wafer (W2), which means they must have the same resistivity. Therefore, chip-to-chip differences are expected to be smaller.

The continuous n-layer sectors for the unirradiated (W2R11) and neutron irradiated samples (W2R1) are shown in Fig. 7.5a and g, respectively. Almost the entire pixel is fully responsive for the unirradiated sample W2R11, except for some areas in the pixel corners. For the neutron-irradiated sample W2R1, the pixel response decreases significantly around the corners and pixel edges. Overall the average pixel

Table 7.3 Calculated average pixel responses for the different samples and designs

Sample	Fluence n_{eq}/cm^2	TID (MRad)	Pixel response [%]		
			Continuous n-layer	Extra deep p-well	n-gap
W2R11	0	0	88.3 ± 2.4	90.5 ± 2.2	90.9 ± 2.2
W2R9	5×10^{14} (p)	66	81.2 ± 2.8	87.6 ± 4.2	88.4 ± 3.8
W2R1	1×10^{15} (n)		75.4 ± 3.8	90.5 ± 2.8	89.0 ± 3.1
W5R9	5×10^{14} (p)	70	80.4 ± 2.8	89.0 ± 2.5	89.3 ± 2.2
W4R9	7×10^{13} (p)	9	78.7 ± 2.6	89.8 ± 2.3	89.9 ± 2.3

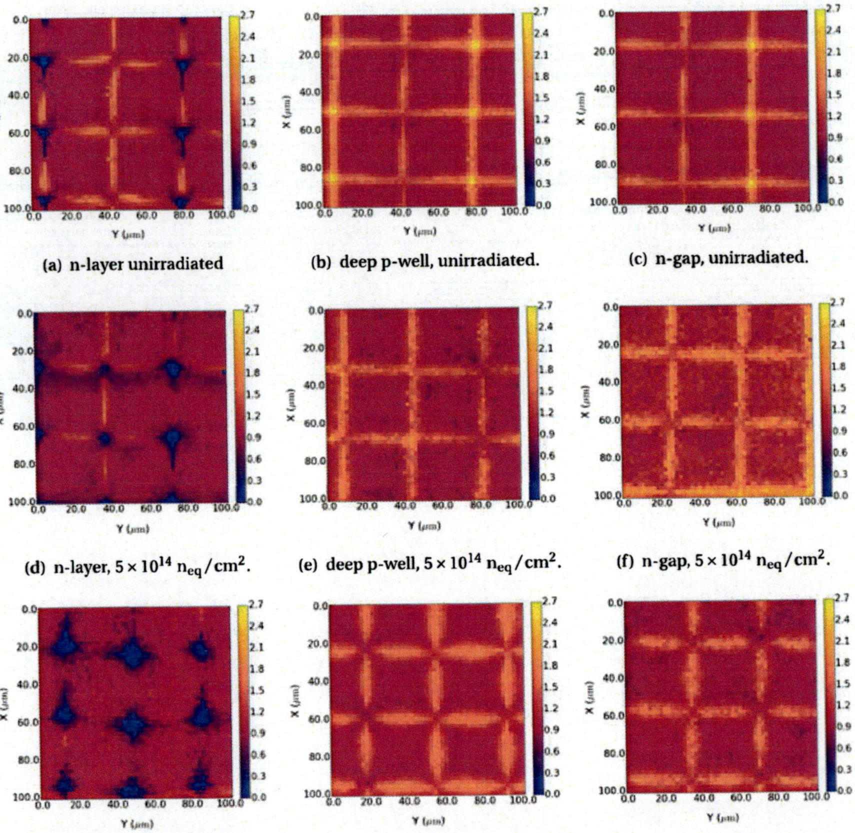

(a) **n-layer unirradiated** **(b)** **deep p-well, unirradiated.** **(c)** **n-gap, unirradiated.**

(d) **n-layer,** 5×10^{14} $\mathbf{n_{eq}/cm^2}$. **(e)** **deep p-well,** 5×10^{14} $\mathbf{n_{eq}/cm^2}$. **(f)** **n-gap,** 5×10^{14} $\mathbf{n_{eq}/cm^2}$.

(g) **n-layer,** 1×10^{15} $\mathbf{n_{eq}/cm^2}$. **(h)** **deep p-well,** 1×10^{15} $\mathbf{n_{eq}/cm^2}$. **(i)** **n-gap,** 1×10^{15} $\mathbf{n_{eq}/cm^2}$.

Fig. 7.5 Pixel response maps for the different samples. For each sample the continuous n-layer, extra deep p-well and n-gap sector are shown. W2R11 is unirradiated, W2R9 is proton-irradiated to 5×10^{14} n_{eq}/cm^2 and 66 Mrad and W2R1 is neutron-irradiated to 1×10^{15} n_{eq}/cm^2

response in the continuous n-sector is reduced at higher fluences (more than 10% after a fluence of 1×10^{15} n_{eq}/cm^2). As previously described, the photon response is not directly comparable to the efficiency determined at proton beam tests. However, the proton efficiency determined at previous a MALTA beam test at SPS [14] shows a similar trend with irradiation. A similar trend was also observed in the mini-MALTA beam test performed at ELSA [2].

Compared to the continuous n-layer, the extra-deep p-well and n-gap designs improve radiation tolerance. There is only a small improvement in the total pixel response value for the unirradiated sample. However, in the pixel response maps (Fig. 7.5b and c), there is no more loss in response at the pixel edges. Moreover, there is almost no decrease in relative response observed with radiation dose for the irradiated samples, and most of the pixel remains fully responsive. Both of the two new tested designs show similar performance. This is in agreement with TCAD simulations [6], which show the same change in electric field lines for both design modifications, and more studies are needed to compare the performance of the two new designs. The other samples studied during the beam test, W5R9 and W4R9, show a similar trend of improved performance with the pixel design modifications.

7.3.3 *Pixel Response as a Function of Bias Voltage*

An additional measurement was performed to compare the pixel response at different bias voltages. The photon response measurements were performed on module W2R1 at a bias voltage of –6 V and –20 V. The resultant response maps for the continuous n-layer sector at each voltage are shown in Fig. 7.6a and b, and the response results are

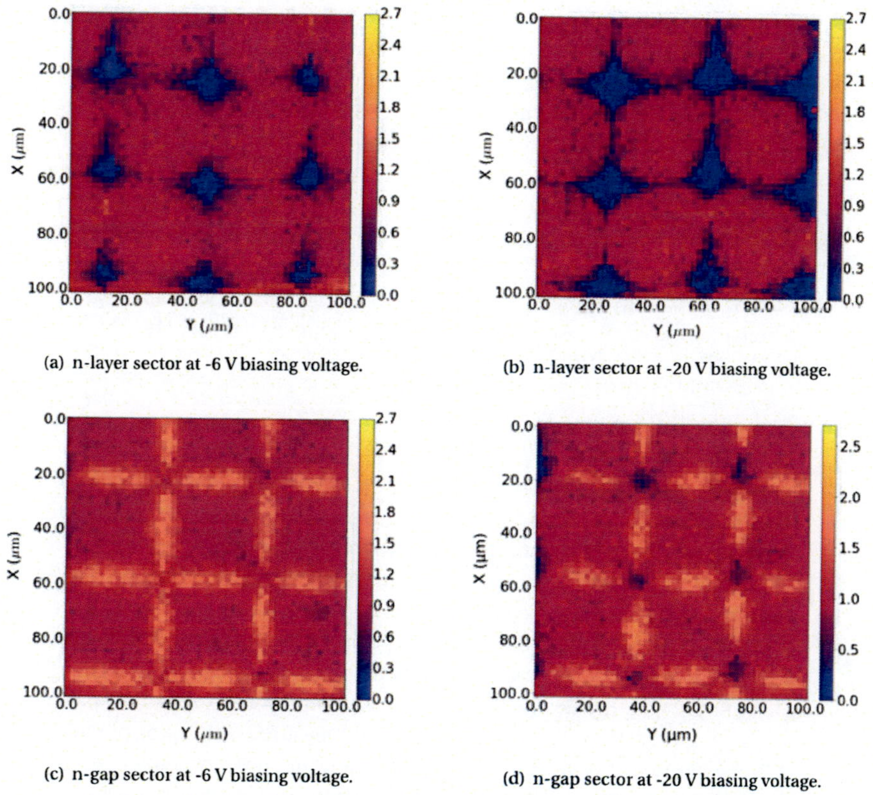

Fig. 7.6 Pixel response maps for continuous n-layer sector and n-gap sector in the W2R1 sample (neutron irradiated to 1×10^{15} n_{eq}/cm^2) at –6 and –20 V biasing voltage

Table 7.4 Average normalised response for the different sectors of W2R1 (neutron irradiated to 1×10^{15} n_{eq}/cm^2) for different biasing voltages

Sample	Bias (V)	Pixel response [%]		
		Continuous n-layer	Extra deep p-well	n-gap
W2R1	–6	76.7 ± 3.8	91.1 ± 3.0	90.0 ± 3.1
	–20	72.2 ± 3.3	86.6 ± 3.9	86.4 ± 2.9

summarised in Table 7.4. The average relative pixel response decreases from 76.7% to 72.2% with increasing bias voltage. This means that the pixel edges become less responsive with respect to the pixel centre and appear sharper. The same effect is seen in the new designs, as shown for the n-gap configuration in Fig. 7.6c and d. Simulations suggest that a higher applied voltage causes a longer drift path and faster drift along the sensor depth [6]. In the pixel centre, the more rapid drift causes better charge collection. However, at the pixel edges, the charge deposited below the deep p-well implant is quickly pushed towards the deep p-well and has to travel almost parallel to the surface. This longer drift path leads to a higher probability of the charge being trapped and not reaching the collection electrode. And as the drift velocity perpendicular to the surface does not increase with higher voltage, more charge carriers are lost, and the normalised response around the pixel edges decreases. A second effect can also have an impact on the relative pixel response: the response is calculated with respect to the charge collected in the pixel centre. With increasing voltage, the depletion depth will get larger, so the total collected charge will increase, and the charge loss at the pixel edges may appear more severe. Therefore, to fully understand this effect, charged particle beam tests are necessary, which can measure absolute efficiency in response to a particle.

7.4 Conclusions and Outlook

This chapter summarised the measurements of a monolithic detector prototype in an X-ray beam test at Diamond Light Source. Monolithic pixel detectors provide an exciting option to reduce material and pixel pitch in pixel detectors at new colliders. Within the ATLAS collaboration, the MALTA and mini-MALTA prototypes were developed. The mini-MALTA prototype includes two design modifications to the substrate for better charge collection at the pixel edges and increased radiation tolerance, improving the design of the MALTA chip. The pixel response of the standard and new structures of mini-MALTA were measured with high precision at a beam test at Diamond Light Source. The two modified designs show significantly better pixel response and less degradation of the performance with irradiation.

Using the data from the X-ray beam test, it is also possible to quantify the charge sharing between pixels and cluster size in the device, with more details described in [15]. While the measurements of pixel response are not directly comparable to efficiency measurements in charged particle beam tests, a conversion between the two quantities is possible with some additional laboratory measurements, as described [12].

As it can be seen from the results presented in this chapter, the design modifications introduced in the mini-MALTA prototype improve the radiation tolerance of the monolithic detector prototype such that it can withstand fluences up to 1×10^{15} n_{eq}/cm^2. Currently, monolithic detectors are not foreseen in the ATLAS ITk upgrade. Still, with further increased radiation tolerance, they might be a feasible option for the replacement of the innermost layers of the ITk, halfway through HL-LHC, which would require radiation tolerance up to 1×10^{16} n_{eq}/cm^2. Efforts to improve the radiation tolerance of MALTA are underway, including tests of different substrates and the improvement of the readout. Beyond the application in ATLAS, monolithic detectors are also an excellent candidate for the detector design for future colliders, such as ILC [16], FCC-ee [17] and CEPC [18].

References

1. Snoeys W et al (2017) A process modification for CMOS monolithic active pixel sensors for enhanced depletion, timing performance and radiation tolerance. Nucl Instrum Methods Phys Res A 871:90–96. ISSN: 0168-9002. https://doi.org/10.1016/j.nima.2017.07.046
2. Dyndal M et al Mini-MALTA: radiation hard pixel designs for small-electrode monolithic CMOS sensors for the High Luminosity LHC. arXiv: arXiv:1909.11987 (physics.insdet)
3. Mager M (2016) ALPIDE, the monolithic active pixel sensor for the ALICE its upgrade. Nucl Instrum Meth A 824. In: Frontier detectors for frontier physics: proceedings of the 13th pisa meeting on advanced detectors, pp 434–438. ISSN: 0168-9002. https://doi.org/10.1016/j.nima.2015.09.057
4. Senyukov S et al (2013) Charged particle detection performances of CMOS pixel sensors produced in a 0.18 μm process with a high resistivity epitaxial layer. Nucl Instrum Methods Phys Res A 730:115–118. https://doi.org/10.1016/j.nima.2013.03.017
5. Pernegger H (2019) Monolithic pixel development in TowerJazz 180 nm CMOS for the outer pixel layers in the ATLAS experiment. Nucl Instrum Methods Phys Res A 924:92–98. ISSN: 0168-9002. https://doi.org/10.1016/j.nima.2018.07.043
6. Munker RM (2018) Simulations of CMOS sensors with a small collection electrode improved for a faster charge-collection and increased radiation tolerance. Pixel
7. Diamond Light Source Ltd, Diamond House, Harwell Science & Innovation Campus, Didcot, Oxfordshire, OX11 0DE
8. Persic A, Nemec T (2007) Ljubljana TRIGA Mark II, 40 years of successful operation
9. Allport P et al (2017) Recent results and experience with the Birmingham MC40 irradiation facility. JINST 12(03). https://doi.org/10.1088/1748-0221/12/03/c03075
10. Displacement damage in silicon. https://rd50.web.cern.ch/NIEL/
11. Nixon MS, Aguado AS (2018) Feature extraction and image processing. Academic Press
12. Metodiev K et al (2020) New method for estimating detector efficiency for charged particles with diamond light source. Nucl Instrum Meth A 982:164573. ISSN: 0168-9002. https://doi.org/10.1016/j.nima.2020.164573
13. Schioppa EJ et al (2019) Measure results of the MALTA monolithic pixel detector. VCI
14. Hiti B et al (2018) Development of the monolithic-MALTA-CMOS sensor for the ATLAS ITK outer pixel layer. In: Topical workshop on electronics for particle physics (TWEPP2018). vol 17, p 21

15. Mironova M et al (2020) Measurement of the relative response of small-electrode CMOS sensors at diamond light source. Nucl Instrum Meth A 956:163381. ISSN: 0168-9002. https://doi.org/10.1016/j.nima.2019.163381
16. Yamada M et al (2018) Development of monolithic pixel detector with SOI technology for the ILC vertex detector. J Instrum 13(01):C01037–C01037. https://doi.org/10.1088/1748-0221/13/01/c01037
17. Koratzinos M (2016) FCC-ee accelerator parameters, performance and limitations. Nucl Part Phys Proc 273-275. In: 37th international conference on high energy physics (ICHEP), pp 2326–2328. ISSN: 2405-6014
18. Guimarães da Costa JB et al (2018) CEPC Conceptual design report: volume 2—physics and detector. Tech Rep 424 pages. arXiv: 1811.10545

Chapter 8
Conclusions and Outlook

The physics program of the ATLAS experiment at the LHC has been hugely successful in discovering and measuring the properties of the Higgs boson. Many production and decay modes of the new particle have been observed, and the observed properties are in good agreement with the Standard Model predictions. However, many open questions remain, and more measurements of the Higgs boson are necessary. For instance, no evidence of its coupling to second-generation quarks has been found yet.

This thesis presented a search for Higgs decays into c-quarks in the VH production mode, using a pp collision dataset of 139 fb^{-1} collected at $\sqrt{s} = 13$ TeV using the ATLAS detector. This search improved on the previous measurement by adding a larger dataset, additional channels, and significant analysis improvements in jet flavour tagging and the estimation of backgrounds. In addition to measuring the $VH(H \rightarrow c\bar{c})$ process, validation measurements of the diboson processes $VW(W \rightarrow cq)$ and $VZ(Z \rightarrow c\bar{c})$ were performed. The diboson measurements showed good agreement with the SM expectation, with observed (expected) significances of 2.6 (2.2) and 3.8 (4.6) for the $VZ(Z \rightarrow c\bar{c})$ and $VW(W \rightarrow cq)$ processes, respectively. The $VH(H \rightarrow c\bar{c})$ signal strength was also found to agree with SM expectation. A 95% CL upper limit of 26 (31) times the predicted cross-section times branching fraction was observed (expected) for the $VH(H \rightarrow c\bar{c})$ process, which significantly improves previous constraints. Additionally, it was possible to measure the Higgs-charm coupling directly for the first time. For the Higgs-charm coupling modifier κ_c an observed (expected) limit of $|\kappa_c| < 8.5$ (12.4) was determined. Finally, in a statistical combination with the existing ATLAS $VH(H \rightarrow b\bar{b})$ measurement, it was possible to derive a constraint on the ratio $|\kappa_c/\kappa_b|$ without assumptions on the total Higgs decay width. The upper limit on this ratio was found to be $|\kappa_c/\kappa_b| < 4.5$ at 95% CL, which suggests that the Higgs boson indeed couples less strongly to c-quarks than to b-quarks, as predicted by the Standard Model.

M. Mironova, *Search for Higgs Boson Decays to Charm Quarks with the ATLAS Experiment and Development of Novel Silicon Pixel Detectors*, Springer Theses,
https://doi.org/10.1007/978-3-031-36220-0_8

The measurement of $H \to c\bar{c}$ decays remains a challenge. It will benefit immensely from a larger dataset, and the upgrade to High-Luminosity LHC (HL-LHC) will provide a dataset of at least 3000 fb^{-1} at $\sqrt{s} = 14$ TeV. With this dataset, an upper limit on the $VH(H \to c\bar{c})$ process of 6.4 times the SM prediction is expected, based on the extrapolation of the current analysis. Even better performance can be expected at the HL-LHC with additional analysis improvements, some already being studied. For instance, using machine learning techniques in the form of boosted decision trees can improve the separation of signal and background and enhance the sensitivity of the analysis by at least 50%. Additional analysis regions, e.g. at high p_T^V, can improve the limit by around 20%, and an improvement in the background modelling uncertainties can also have a significant impact. Finally, better capabilities for identifying c-jets are crucial for this analysis. These are expected to improve due to improved flavour tagging machine learning algorithms and the better resolution of the ATLAS inner detector at HL-LHC.

Significant modifications are needed to the LHC and its experiments to collect the larger dataset at HL-LHC. For instance, the ATLAS inner detector will be replaced with an all-silicon inner tracker (ITk). This detector must have a significantly improved radiation tolerance, better spatial resolution and faster readout to cope with the higher instantaneous luminosity and pile-up expected at the HL-LHC. The pixel readout chips are particularly susceptible to total ionising dose (TID) effects. In the innermost layers of ITk, the detectors are expected to receive TID doses of up to 1 Grad after 2000 fb^{-1} of delivered luminosity, after which they will be replaced. The ATLAS pixel readout chip was designed by the RD53 collaboration to cope with these HL-LHC requirements.

In this thesis, the latest chip prototype, ITkPixV1, was extensively tested in irradiations using X-rays and protons. Different chip components were tested, such as the digital logic, the analogue front-end and the chip powering circuit. The radiation damage to the digital logic was characterised in measurements of gate delay increase using dedicated ring oscillators. The ring oscillators were found to be radiation tolerant up to total doses of at least 2.3 Grad at a dose rate of 4 Mrad/h. However, a significant dose rate dependence was observed, with more damage observed at lower dose rates. Around 2–2.5 times higher damage was found at a low dose rate. For the gates used in the chip digital logic, this corresponds to an expected gate delay increase of around 100% after 1 Grad of low dose rate irradiation, which is well within the required margins.

The ITkPixV1 analogue front-end was found to be radiation tolerant up to total doses of at least 1 Grad. Most of the damage to the analogue front-end occurred at the start of the irradiation. No dose-rate dependence was observed for the radiation damage to the larger transistors used in the analogue front-end. The chip powering circuit (SLDO) was also characterised during irradiation, and no start-up issues were found after irradiation to 1 Grad. Finally, the behaviour of the ITkPixV1 chip was studied in pixel modules after irradiation with 27 MeV protons. The performance of the analogue front-end deteriorated at high leakage currents (10 nA/pixel) and low temperatures (<-20 °C) but could be recovered with the use of the leakage current compensation circuit in the chip. Overall, the ATLAS readout chip shows excellent radiation tolerance and is expected to survive the lifetime of ITk in terms of radiation damage.

Beyond the planned ITk upgrade, new detector technologies provide a possibility to improve our detector performance and physics measurements even further. Monolithic pixel detectors offer a promising opportunity to reduce pixel pitch, detector material and production costs. However, the technology is less radiation tolerant than hybrid pixel detectors. This thesis discussed improvements to a monolithic detector prototype in 180 nm TowerJazz technology. The mini-MALTA prototype included two modifications to the standard detector design, characterised by an X-ray beam test at Diamond Light Source with high precision. The measurement provided high-resolution maps of the pixel response and indicated that the two process modifications significantly improve the radiation tolerance of the prototype, up to fluences of $1 \times 10^{15}\ n_{eq}/cm^2$.

In summary, with the expected increased dataset collected by the ATLAS detector in the coming years and many ongoing efforts to improve our detectors, exciting times are ahead for measurements of the Higgs boson and its properties.